PRIVACY IN A PUBLIC SOCIETY

PRIVACY IN A PUBLIC SOCIETY

Human Rights in Conflict

RICHARD F. HIXSON

New York   Oxford
OXFORD UNIVERSITY PRESS
1987

Oxford University Press

Oxford   New York   Toronto
Delhi   Bombay   Calcutta   Madras   Karachi
Petaling Jaya   Singapore   Hong Kong   Tokyo
Nairobi   Dar es Salaam   Cape Town
Melbourne   Auckland

and associated companies in
Beirut   Berlin   Ibadan   Nicosia

Copyright © 1987 by Richard F. Hixson

Published by Oxford University Press, Inc.,
200 Madison Avenue, New York, New York 10016

Oxford is a registered trademark of Oxford University Press

Library of Congress Cataloging-in-Publication Data
Hixson, Richard F.
Privacy in a public society.
Includes index.
1. Privacy, Right of—United States—History.
I. Title.
KF1262.H55   1987   342.73'0858   86-18161
347.302858
ISBN 0-19-504292-1

2 4 6 8 9 7 5 3 1
Printed in the United States of America

For Todd, Sommer, and especially Cynthia,
whose strong sense of community
binds our private worlds

Acknowledgments

In the preparation of this book I have depended upon the advice and assistance of many persons, not the least of whom are my students and colleagues at Rutgers University, my home-away-from-home for more than twenty-five years. President Edward J. Bloustein's seminal work on privacy stimulated my own thinking toward a different stance. Dean Richard W. Budd has always encouraged my interest in historical scholarship. Jerome L. Aumente allowed me to participate in a number of press-bar-bench workshops at the Journalism Resources Institute. David B. Sachsman thought my "anti-privacy" thesis was worth the shelter he provided from departmental responsibilities. I especially want to thank Elisa Bildner, a colleague and lawyer, for her scrutiny of the manuscript when it really counted at deadline, and Peter Ehrenhaus and Richard Morris for their tolerance when I needed help with philosophical issues. Richard Quaintance and John Chambers also expanded my horizons. Also at Rutgers, Todd Hunt, a good and faithful friend, purged my writing as best he could, and Lloyd C. Gardner introduced me to Sheldon Meyer at Oxford, who responded immediately to the project at its beginning and put me in the good charge of Joellyn Ausanka, Rachel Toor, and Stephanie Sakson-Ford. Thanks again to each.

I am much indebted to the librarians who acquainted me with their collections and bore with my questions, especially those at the Alexander Library at Rutgers, the New Jersey State Law Library in Trenton, and the Hunterdon County Law Library in Flemington. Steven B. Frakt of the New Jersey Division of Legislative Information and Research filled me in on how one state deals with citizen privacy. I wish also to thank George Hoerrner, a senior partner in the law firm of Gebhardt and Kiefer in Clinton, New Jersey, for giving me access to the firm's library, and to attorneys Albert D. Rylak and Robert DeChellis for the many hours over coffee at Nick's Towne Restaurant. Thomas Cafferty, attorney for the New Jersey Press Association, has advised me on communication law for several years.

I could never have finished the book without the personal and professional commitment of Alfred McClung Lee and Brent D. Ruben. Al Lee, prominent sociologist and dear friend, endured every word and challenged every paragraph. He drew my attention to the importance of social traditions, or folkways, in the way Americans have perceived their personal privacy and collective survival over time. He is correct in noting that cultural contrasts have far-reaching consequences for the way individuals and groups define privacy.

Nor can I say enough to express my gratitude to Professor Ruben, who also read every word. He took valuable time away from his own research to urge important changes in the narrative that would make the volume appealing to the layperson as well as to the specialist. I wish also to thank him for permission to reprint portions of Chapter 7, which appeared as "Whose Life Is It, Anyway? Information as Property," in *Information and Behavior* (New Brunswick and Oxford: Transaction Books, 1985). Professor Ruben is the founding editor of that important new series.

Charles Nutt, executive editor of the *Courier-News* of Bridgewater, New Jersey, and that newspaper's chief editorial writer, Robert Comey, gave me the forum of its Op-Ed Page on two occasions. I appreciate their interest in academic research.

The extent of my family's contribution is immeasurable. For their patience and understanding I thank my wife, Cynthia, and our children, Todd and Nancy Sommer. Our family now includes Tracy Wheeler, who, with Todd, read an early draft and said, in chorus, "Lighten up!" Recent college graduates, they know too well the burden of having to read unnecessary pages. Cynthia, who trained our Labrador while I read books, is especially glad to see the project in this form. To each, my deepest thanks.

Clinton, New Jersey   R. F. H.
June 1986

Contents

Introduction

Privacy is akin to the tried-and-true forces that irresistibly impel individuals toward association: hunger, love, vanity, and fear. All are common and recognizable human impulses that begin in isolation but soon become socializing forces that compel necessary cooperation. But, if allowed to fester long in isolation, these impulses may turn into forces antagonistic to society. Inevitably, society advances or regresses as individuals choose either a private existence or a life of public coexistence.

Privacy begets conservatism, retrogression, and stagnation, the social scientist James G. Leyburn observed many years ago, but more privacy or further isolation is not the answer. Any such tendency is certain to retard the progress of society, especially a modern complex society whose members are by force of circumstance interdependent. Leyburn wrote: "Where selfishness in the individual or ethnocentrism in the group exceeds the willingness to make concessions to one's associates, there is an inevitable tendency toward disintegration of the bond." Indeed, as Octavio Paz implied in *The Labyrinth of Solitude*, private and public are at once complementary and contradictory.

When it comes to an assessment of privacy as a "right"—"the right to be let alone"—the reader will soon discern that privacy, although a natural human desire, is not a natural right in the way John Locke originally interpreted the term. Locke said that man possessed certain "natural" or "God-given" rights, and that a human right thus defined was a power the individual retained over against the state. This book argues that the right to privacy is a power exercised in the name of the state, or, more directly, the community. Privacy is a right created and bestowed by others, or else it is not a right at all.

This is what William Graham Sumner meant when he documented early in this century that rights are the rules of mutual give and take and are imposed in order to sustain peace and group strength. They are products of the folkways. "They are reflections on, and generalizations from, the experience of pleasure and pain which is won in efforts to carry on the struggle for existence under actual life conditions," Sumner wrote. To Thomas Jefferson, who did not subscribe to the Lockean notion of an original "sovereignty" over oneself, a right was directed toward others. It is based upon the principle of sociability, of benevolence, not selfishness. Arguably, the right to privacy is a societal bequeathal, not an individual possession. "The test is public good," writes Garry Wills on the Jeffersonian ideal.

When I first wrote lectures for a university course called "Communication Law," I gave only routine attention to privacy, as compared to what I thought were more important—and, frankly, more interesting—topics. Today, the course risks being dominated by privacy. It has become an important and interesting topic. This book is the result of that transformation, a change in my own perception of privacy as well as an alteration in society's concept of what is private. And, although I admit to the value of privacy, I think that our mutual concern is overdrawn and problematic.

We have traveled far from our earliest concepts of privacy as solitude, seclusion, and solitariness. People today are jealous of their privacy. They fear the loss of personal privacy, more so since the advent of impersonal computerized information banks. Once more, they regard privacy as a natural right, for only God appears to be on their side. It may be only a perceived problem, but it is real nonetheless. The right to privacy is more than a legal concept; it is also a religious tenet, a cult, a fetish. But, as the early chapters show, it is at the center of basic American individualism. Privacy may not be a natural right, but it is certainly historical, institutional, and empirical. It is now also legal. A. W. Arndt, the Australian writer, captured privacy's history when he said nearly forty years ago,

> As privacy came more and more to be regarded as a natural right, so publicity — the pillory — went out of use as a means of enforcing social behavior. Some time between the middle ages and the nineteenth century, the individual appears to have obtained two substantial advantages at the expense of society: the right to keep his own affairs to himself and the abandonment of publicity as a means of ensuring conformity to social codes.

The cult of privacy (Arndt's phrase) is a byproduct of individual competitiveness.

Privacy rests upon an individualist concept of society, not only in the beneficial sense of society wherein the goal is the welfare of all citizens, but in the more specific sense of "each for himself and the devil take the hindmost!" Or, as Anthony Burgess, the British writer, observed during one of his frequent visits, "Americans don't share things in common; they have their own things." Ironically, Americans have a lot in common, even, I would argue, their privacy, but because they equate personal privacy with private ownership they are witnessing the decay of their commonality and their public amenities. Privacy has its price!

Numerous other writers have sought to understand the relationship of private and public in American life. This particular contribution to the growing literature is the role law-making has played in preserving the best of both worlds. In tracing the *legal* history of privacy, I have selected as primary sources the court cases that best depict the continuing struggle over the meaning of private and public, the jurisprudential debate over individual autonomy and collective welfare, between the person and the state, the individual and the community. My secondary sources are the works of historians, social scientists, legal scholars, and journalists, whose findings and reportage elucidate the conditions that have generated laws affecting privacy.

PRIVACY IN A PUBLIC SOCIETY

The Unknown Citizen
(To JS/o7/M/378)
This Marble Monument Is Erected by the State

He was found by the Bureau of Statistics to be
One against whom there was no official complaint,
And all the reports on his conduct agree
That, in the modern sense of an old-fashioned word, he was a saint,
For in everything he did he served the Greater Community.
Except for the War till the day he retired
He worked in a factory and never got fired,
But satisfied his employers, Fudge Motors Inc.
Yet he wasn't a scab or odd in his views,
For his Union reports that he paid his dues,
(Our report on his Union shows it was sound)
And our Social Psychology workers found
That he was popular with his mates and liked a drink.
The Press are convinced that he bought a paper every day
And that his reactions to advertisements were normal in every way.
Policies taken out in his name prove that he was fully insured,
And his Health-card shows he was once in hospital but left it cured.
Both Producers Research and High-Grade Living declare
He was fully sensible to the advantages of the Instalment Plan
And had everything necessary to the Modern Man,
A phonograph, a radio, a car and a frigidaire.
Our researchers into Public Opinion are content
That he held the proper opinions for the time of year;
When there was peace, he was for peace; when there was war, he went.
He was married and added five children to the population,
Which our Eugenist says was the right number for a parent of his generation,
And our teachers report that he never interfered with their education.
Was he free? Was he happy? The question is absurd:
Had anything been wrong, we should certainly have heard.

March 1939
W. H. AUDEN

1

In Search of Solitude

Privacy, in practice as well as in concept, has long been a central part of mankind's history. The writer of Genesis describes the moment when Adam and Eve discovered what it meant to be private: "And the eyes of them both were opened, and they knew that they were naked; and they sewed fig-leaves together, and made themselves aprons."[1] They were said to know instinctively, but not without God's help, both the feeling of privacy and the loss of it. Genesis also records the awarding of God's first gift to man besides the gift of life itself—the right to be reticent before the eyes of each other. "And Jehovah God made for Adam and for his wife coats of skins, and clothed them."[2] This passage may be extended as a sign to our own time, according to John Curtis Raines, "a warning written over man's interaction with his fellows of the right of self-defense against the imperialism of relentless inquiry, against all watching that would everywhere follow, probe, and hold a person within its sovereign gaze."[3] Even God, as portrayed in the Old Testament, refused the power of gazing upon the privacy of his children, and thus sought to establish the civil right we now associate with human personal privacy.

John Milton went beyond the mere personal in privacy to the poet's version of a more universal concept. *Paradise Lost* is about the loss of Eden, the loss of innocence, and the loss of privacy. Arnold Stein's analysis of Milton's epic message is insightful for what it reveals about the private life:

> Related (to the known cultural history of *Gan Eden*) is the occasional human yearning for peace and quiet, for an impossibly ideal Arcadia where wilderness may be thought paradise enow; or there is the desire to withdraw from complexity, from 'the fury and mire of human veins,' to withdraw even, especially, from the self and its involvements. . . .[4]

That is what the critic sees in the poet's vision, and Milton's version also suggests the terrible irony in man's ageless search for solitude and seclusion, the paradise that is inevitably lost and the privacy that is so temporary and nearly always invaded. There is first the yearning, which is probably genetic and God-given; there is the reality, too, which is often beyond the individual's control:

>   and next to Life
> Our Death the Tree of Knowledge grew fast by,
> Knowledge of Good bought dear by knowing ill.[5]

Little did Adam and Eve realize at the time that their "happy rural seat," surrounded by a high "verdurous wall," would soon be invaded, first by Satan as "leviathan-explorer-merchant-griffin-vulture-scout-wolf-cormorant," and then by God's own messenger, the Archangel Michael, who, in the final scene of *Paradise Lost*, instructs them to leave:

> The World was all before them, where to choose
> Thir place of rest, and Providence thir guide:
> They hand in hand with wandring steps and slow,
> Through *Eden* took their solitarie way.[6]

Thus concludes this account of man's first experience with privacy and its loss.

Later on, the writer of Genesis records another brief encounter, emphasizing this time man's natural yearning for privacy, his need to go unobserved:

And Noah began to be a husbandman, and planted a vineyard: and he drank of the wine, and was drunken; and he was uncovered within his tent. And Ham, the father of Canaan, saw the nakedness of his father, and told his two brethren without. And Shem and Japheth took a garment, and laid it upon both their shoulders, and went backward, and covered the nakedness of their father; and their faces were backward, and they saw not their father's nakedness."[7]

The chronicler makes no mention of the brothers' *desire* to intrude; we are told only that they sensed an impropriety, and, although the one son had in fact intruded upon his father's solitude, the three instinctively understood their collective duty. To stretch the metaphor, this Biblical reference illustrates a recurring aspect of personal privacy—the conflict between the subjective desire for solitude and seclusion and the objective need to depend upon others. We learn from Milton's version of the Fall and from the collective action of Noah's sons that privacy is not a solitary event, for others (no matter how many) are nearly always involved. Adam and Eve discovered the ultimate privacy but forswore their obligation to keep it; Noah's sons learned from Adam and Eve's mistake.

The substantive themes running throughout the history of privacy are similar to those encountered in Biblical times. Privacy as a right against authority was also integral to early Hebrew culture. Classical Greece and ancient China also took pains to formulate social norms for protecting individual autonomy from the rest of society. Yet, the dominant value was not privacy, nor can it be in any society, as Barrington Moore, Jr., learned from his study of the three civilizations. "Man has to live in society, and social concerns have to take precedence," he concluded. In ancient Athens and in China, private life was generally seen as a manifestation of antisocial behavior, although both cultures recognized the private as well as the public realm. The Hebrews, however, made little distinction between private and public, and God alone, it seems, enjoyed privacy. Although all three civilizations were more concerned about social, or public, matters, not all public concerns took precedence. Sometimes, but not routinely, individual privacy was tolerated—even respected.

In Moore's view, such early concepts of privacy marked the beginning of Western civilization's attempt to achieve a balance between personal privacy and public authority. He also notes that without respect for human dignity, then as now, there can be no individual autonomy in the sense of either seclusion or protection against authority. Moreover, without democracy private rights are either stunted or absent.[8]

With the spread of Christianity, the conflict between privacy and authority became more acute, epitomized, and brought to a head by the practice of the church-directed confessional. Such intrusions into private matters created anxiety and resentment among increasingly literate parishioners. The role of the individual in matters of religion continued to plague the church and was of major significance in the Reformation, which, according to Moore, "brought about a new emphasis on personal faith and individual conscience that has made a contribution to conceptions of individual moral autonomy in the modern world."[9]

Private interests in the Middle Ages were seldom honored, for monarchies and churchmen were preoccupied with constant battles for power. It remained for the triumph of commerce and industry, Moore writes, for a less parochial bourgeoisie to generate major new contributions to the theory and practice of both privacy and private rights. For most of its long history, in fact, privacy has seldom been defined as a phenomenon separate from ordinary social custom or personal behavior. People simply knew what it was. It was not until the latter part of the nineteenth century that writers began to think of privacy in terms of legal protection.[10]

Raymond Williams, the British scholar who has written extensively on individual relationships in a community setting, traces the term "privacy" to its root, *private* and its Latin forebears, *privatus* and *privare*. The former means withdrawn from public life and the latter to bereave or deprive. In the fourteenth century, private was applied to withdrawn religious orders, "where the action was voluntary," and from the fifteenth century to persons not holding public or official position or rank, as, for example, "private soldier" and "private member" (in Parliament).

Eventually, private acquired the sense of secret and concealed, both in politics and in the sexual senses of "private parts." Williams notes a crucial moment of transition when the term acquired a conventional use opposite that of "public," as in "private house," "private education," "private view," "private club," and "private property." In virtually all these uses, Williams discerns, the primary sense was one of *privilege*. "The limited access or participation was seen not as deprivation but as advantage."[11] Private acquired this favorable meaning, a sense of exclusivity, from about the sixteenth century and running through the nineteenth.

"Privation," according to Williams, retained its old sense of being deprived, and "privateer" its sense of seizing the property of others, as in the original "private man of war." *Privilege*, meanwhile, went with "private" (*privilegium* in Latin, a law or ruling in favor or against an individual) and implied a special advantage or benefit. In this regard, it is interesting to note that Karl Marx categorized private ownership, private enterprise, and private greed as "naked self-interest." As is discussed later, private or privacy carries with it the notion of privilege, both in the sense of the privileged class that enjoys more privacy and in the belief that privacy is a condition worthy of legal protection, a lawful immunity from officialdom.

Still another important movement replaced "withdrawal" and "seclusion" with the senses of private as "independence" and "intimacy," as in "the privits of my hart and consciance," a line from the mid-sixteenth century. Thus, "private friends," for example, implied not only intimacy but a kind of special or privileged intimacy. Williams records how, in the seventeenth and eighteenth centuries, seclusion in the sense of a quiet life was valued as "privacy," and, he writes, "this developed beyond the sense of solitude to the senses of decent and dignified withdrawal and of the 'privacy of my family and friends,' and beyond these to the generalized values of 'private life.' " Obviously, privacy in these senses was related to corresponding changes in notions of "individual" and "family."

Williams believes that "private life" has retained its old sense, as distinct from "public life," but it is the progressive association of "private" with *personal*, as very favorable terms, that now appears predominant. The term is still frequently used in unfavorable ways, however, as in "private profit" or "private advantage" or as in the Marxian sense mentioned earlier. "Private enterprise" receives mixed reviews, depending upon its negative or affirmative context. Williams concludes his history on a somewhat political note:

> Private . . . in its positive senses is a record of the legitimation of a bourgeois view of life: the ultimate generalized privilege, however abstract in practice, of seclusion and protection from others (*the public*); a lack of accountability to 'them'; and of related gains in closeness and comfort of these general kinds. As such, and especially in the senses of the rights of the *individual* (to his "private life" or, from a quite different tradition, to his *civil liberties*) and of the valued intimacy of *family* and friends, it has been widely adopted outside the strict bourgeois viewpoint. This is the real reason for its current complexity.[12]

All of the ancient notions of solitude and seclusion were evident in early America as well, where privacy was a serious matter but not one of dominant concern. Personal privacy competed with the needs of defense, what Barrington Moore, Jr., perceives as "collective survival," and political organization, religion, and just plain sociability. The Puritan settlements were a collective venture, according to David H. Flaherty, and they were based upon a belief in community that required the population to band together for protection and mutual encouragement. "Privacy took second place to other values in the location of homes until Puritan communitarian ideals gradually disintegrated in the face of New World conditions," writes Flaherty.[13] In a phrase, privacy and elbowroom went hand in hand.

While colonial officials, such as legislators and churchmen, lamented the spread of population to the countryside, beyond the compact towns designed and encouraged with communal life in mind, the citizens, however, had begun to acquire an "insatiable desire" for more land around their dwellings. In Flaherty's view, the case of access to ownership or possession of land in the New World furnished a secure base for the enjoyment of privacy. Because of the increasing distance between homes, particularly farm homes, physical privacy became a characteristic of everyday life. "The ordinary colonial family could enjoy intimacy, as well as an almost automatic degree of solitude," Flaherty writes.[14]

Ironically, however, as the larger New England towns (i.e., Boston, Newport, New Haven, New London) grew rapidly in population and physical privacy was impeded, the emerging urban centers provided the "protective anonymity" that could less readily be found in the smaller towns. No doubt the impersonal nature of big city living, then as now, contributed to personal privacy at least, if not spatial seclusion. Cultural values, as well as economic status, influenced the size of homes persons wanted—including a desire for more privacy within the house itself.

The colonial household between 1700 and 1740 was necessarily a rather basic enterprise, even in the growing seaport towns, with no running water, only passable lighting, no refrigeration or screening, and minimal privacy. Few houses had corridors or anything else allowing access to one bedroom without going through another. Overhead the floorboards of the lofts where older children and sometimes servants slept usually had wide cracks between them, which let sound pass freely but also allowed some heat from the first-floor fireplace to rise upwards.[15]

In general more and more physical privacy was available

within colonial homes with the passage of time, Flaherty says.[16] A significant event was the introduction of hallways in larger eighteenth-century homes, an addition that heightened privacy by eliminating the need to pass through rooms to get to other parts of the house. Soundproofing, however, was generally unsatisfactory, except for families that could afford thicker plaster-like walls. Flaherty notes: "Changes in the impact of the Puritan movement on society as a whole encouraged the growth of conspicuous consumption, while the attractiveness of forced communal living in a nuclear center declined."[17]

The home was the primary place of privacy for the colonists. In the construction of their dwellings, they sought to strike a balance between intimacy for the family as a unit and solitude for individual members of the household. Increased wealth was a major factor in the creation of private space. The poor lived in physically cramped quarters, but this is as common today as it was then. Children probably enjoyed as much, if not more, privacy than their elders because they were ignored by busy parents.[18] This condition seems not to have changed over time, for today's so-called "latch-key" children enjoy enviable—by peer assessment—private lives.

The desire for privacy in daily life then, as now, was also a by-product of a person's work and the work-place environment. A factory worker today, for example, is likely to seek solitude during his or her leisure hours, but among the colonists only those individuals engaged in intellectual pursuits, such as clergymen, were overly concerned with solitude. Flaherty's study is again helpful:

> The availability of solitude within the home outside of working hours was not a pressing issue for the colonists; their need for individual privacy in this sphere was less than that felt in industrialized societies.

Leisure time was for socializing, not for privacy.

Privacy and sexual activity go hand in hand, and, from what we can tell, always have. In fact, the ideal, in terms of a desire for privacy during sexual contact, quite clearly transcends Western culture, as Barrington Moore, Jr., shows in his review of anthropological evidence.[19] Nearly all societies, primitive as well as modern, have sought privacy for sexual relations, and early America was no exception. Crowded quarters and the lack of insulation in New England homes no doubt challenged the imagination of couples, but the expectation of sexual privacy was always present.

Most rooms, upstairs or down, contained beds to accommodate at least two persons. Frequently the formally designated bedroom was a several-beds dormitory. And, as J. C. Furnas writes amusingly: "Trundle beds pulled out from underneath accommodated still more sleepers, typically the smaller children who must have become early acquainted with the facts of life as their parents demonstrated them on squeaky frameworks of wood and rope."[20] Flaherty relates numerous instances where the desire for sex was accommodated by some degree of seclusion, in wooded areas or barns, if bedrooms were unavailable or too crowded. The home was the usual place for both married and unmarried couples, but, as the saying goes, where there was a will the desiring couple found a way. Sometimes the occasion was embarrassing. In Sutton, Massachusetts, in 1760, George Gould went to the home of his friend Edward Holman "and knockt at the door, but no body coming, he pushed the door open," much to the surprise of Mrs. Holman and her lover, who were just getting out of bed. Most visitors were less headlong and respected privacy, however.[21]

Plymouth Plantation was the scene of one of the first recorded instances of what today we would call "invasion of privacy" by the government. In 1624, only four years after the Pilgrims had landed and endured their first American winter, Governor William Bradford learned of a plot against the leadership of the small colony. He had intercepted several incriminating letters written by two newcomers and sent to friends in England. Bradford summoned the men before an assembly, confronted them with the conspiracy, and when they denied the accusation the governor produced the letters and "caused them to be read before all the people." The men expressed outrage that their private correspondence had been intercepted, but they "were silent, and would not say a word" when Bradford asked if they thought the magistrate had "done evil" in opening the missives.[22]

This episode illustrates privacy at its most personal and primitive level, but it also explains one of the reasons the defendants remained silent when given the chance to voice their indignation. They had no legal leg to stand on, for privacy as a part of English common law did not exist in the seventeenth century. Neither was the concept to be found in the Biblical law of the Puritans, nor in the mixture of both traditions that was prevalent in colonial Massachusetts. In later years the American settlers would enlarge upon basic individual rights in their colonial charters, and when the Declaration of Independence was written, and eventually the Constitution, they spelled out such rights as freedom of religion, freedom of speech, freedom of conscience, of assembly, and freedom from unreasonable search and seizure. However, it would be more than a century before the additional right to privacy emerged as a support for these original, and more basic, freedoms. Even at that, the beginnings of the legalization process, the invisible hand of the future was apparent in the Reverend Ebenezer Devotion's sermon in 1753:

> So different are the private Interests of Men, so various and headstrong their Passions and Lusts, that good Laws and a strict Execution of them, will probably bring upon you the Odium of many.[23]

As we see later, modern privacy "rights" have indeed proliferated. Privacy has become the trendy byword for a number of demands, and the courts have expounded and extended protection of privacy as nearly the equivalent of all civic order.

The eighteenth-century home, meanwhile, was both a place of shelter and a retreat from the outside world, much as it is in the twentieth century. When the Massachusetts Excise Bill of 1754 required each homeowner to tell the tax collector how much rum his household had consumed during the year, pamphleteers denounced the regulation as a breach of domestic privacy. In the words of one writer,

> It is essential to the English Constitution, that a Man should be safe in his own House; his House is commonly called his Castle, which the Law will not permit even a sheriff to enter into, but by his own Consent, unless in criminal cases.[24]

Not only did a citizen believe he had a right to privacy in his home, recognized in common law by that time, but he also believed he had the privilege, also enforced by law, to repel anyone who challenged that right. As part of his defense against general search warrants in 1761, James Otis, the prominent Boston attorney, said:

> Now one of the most essential branches of English liberty, is the freedom of one's house. A man's house is his castle; and while he is quiet he is as well guarded as a prince in his castle.[25]

If strangers happened by or neighbors intruded, the colonists reacted politely but firmly. The native Indians were told to first "knocke att the dore, and after leave given, to come in (and not otherwise)," but the rules applied to everyone. Eavesdroppers and Peeping Toms were often prosecuted for invading domestic privacy. What is interesting, hospitality was more open in the seventeenth century than in the eighteenth, when, as Flaherty discovered, colonists started to limit their availability at home. Flaherty says that this was more prevalent in the settled parts of the country and was especially the case among the upper classes. "Their homes had never been so open, in part because a stranger without a letter of introduction would approach a less imposing home," writes Flaherty.[26] Thus, with wealth and property also came the privilege, in addition to the developing right, of personal privacy.

Neighbors, especially those who lived close by had great surveillance powers, because they could easily view any movement next door and they were also close enough to overhear actual conversation. "Disturbances of the neighborhood," as they

were called, were routinely presented in court and the per-
petrators dealt with. In the mid-eighteenth century, the Reverend Andrew Eliot of Boston admonished his flock to be "faithful
Monitors" of their neighbors "when they have done amiss,"
implying that religion then, as today, has the responsibility, if
not the right, to interfere with an individual's privacy as part of
a higher moral code.

"Individuals usually protested neighborhood surveillance
without recourse to the courts," writes Flaherty, "probably beginning with a conversational hint, and ultimately resorting to
fisticuffs and gunfire if their privacy continued to be invaded."[27]
Such harsh methods, by today's standards, presumably were
effective, tempting one to conclude that law and justice do not
always carry the same meaning, as some legal conservatives
maintain. Flaherty's observation is particularly cogent:

> Although colonial New England was in general a democratic soci-
> ety . . . the concept of deference, which pervaded all aspects of
> life, modified any elements of antipathy to privacy implicit in the
> notion of democracy. This was unlike the situation in late-nine-
> teenth-century America when a popular interpretation of the
> meaning of democracy sometimes associated a conscious search
> for privacy with aristocratic tendencies. The colonial sense of def-
> erence stimulated a respect for other persons, particularly one's
> betters, and for their privacy.[28]

But the idea of deference also extended to anyone the guarantee—and eventually the legal right—to seek personal privacy if
he or she so desired. Privacy and democracy were frequently at
odds, competing claims that recur throughout this study to confound the search for, let alone the legal protection of, personal
solitude and seclusion. What is suggested here is that, during an
earlier time in American history, privacy was more a matter of
honor than of law. Later, the reverse of that practice appears to
have become the norm, as is evident in later chapters.

Flaherty, in his study of colonial privacy, takes heart from the
fact that, although people had some trouble being alone in close
families or going unrecognized in friendly neighborhoods, they

believed that the benefits of a warm atmosphere usually outweighed any diminution of personal privacy. Today, there
seems to be as much fear of intimacy as there is fear of loss of
privacy. The colonists neither feared intimacy nor viewed it as
an inevitable threat to their privacy. They always had the option
to withdraw temporarily for a "period of retirement."[29] People
may not have enjoyed a perfect balance between privacy and
society's competing claims, but such dilemmas, then as now,
are inherent in the basic concern for privacy. The issue is no
more complex today than it was during the country's formative
years, but the *ways* privacy is violated are substantially different
and, perhaps, more threatening. Typical of the many alarums in
the twentieth century is the following:

> This invasion is almost inevitable given . . . the development of
> sophisticated electronic devices such as directional microphones,
> powerful miniature listening devices, telephoto lenses and the all-
> pervasive computer with its power to store and retrieve the min-
> utiae of our lives. The problem is exacerbated by the power of the
> mass media to disseminate widely information about individuals,
> including their physical images.[30]

Besides visitors and neighbors, the size of colonial communities
affected privacy, as did the range and number of personal ac-
quaintances, the prevalence of gossip, and facilities and technology for spreading information. Anonymity and solitude
were always possible but harder to achieve in the larger towns.
Such important aspects of privacy depended upon the amount
of support a community gave to personal desires. And such
disadvantages as bigness were no doubt overcome by the mutually supportive nature of the community.

Of some importance, too, was the advent and spread of newspapers in the eighteenth century, a means of communication
that probably improved the reliability of information that had
heretofore been hearsay. Information increasingly came from
public and other official sources, rather than from friends and
neighbors. On the other hand, the newspapers tended to gener-

ate a hunger for more information and thus presented another kind of challenge to personal privacy. As Flaherty surmises:

> The concentration on local affairs presented a threat to the privacy of any local person whose activities became even remotely worthy of attention. This was one of the negative aspects of community life for which some of its more pleasurable characteristics had to compensate.[31]

It is reasonable to assume that because of newspapers the colonists developed an inquisitiveness that reached beyond their own towns, therefore, the press, in response, eventually developed aggressiveness.

In many ways, it is a short distance between the Garden of Eden, the first recorded instance of man's duel with himself and concert with others, and colonial America, another beautiful wilderness in which the private self sought both refuge and collective identity. As the philosopher Paul Weiss writes: "A human corpse is initially a whole and later an aggregate but, through society's conditioning, is made the object of respect and even reverence by others."[32] Or, as John Curtis Raines sees the relationship, the dual nature of our lives: "We seek privacy because we are essentially, and inevitably, private persons who inhabit separate bodies. But even as private persons we need also to disclose ourselves, share our secrets, seek the comfort of human companionship, and in other ways depart our solitude and seclusion."[33]

The novelist provides still another dimension to the human paradox. Anthony Keating, the protagonist in Margaret Drabble's marvelous depiction of modern urban decay and the vicissitudes of country life, *The Ice Age* (1977), muses over his decision to leave London:

> Not only had the boom turned into a slump: his life, which had recently been far too full, had suddenly become extraordinarily empty. He would have to learn to cope with solitude. It had become the new problem. Occasionally he felt an urge to drive down to London, just to see what was happening, although he knew he

could do nothing useful: part of him missed the anxiety, the tension, the racket. But he would train himself to stare at stones and trees. It was a longer-termed insurance. If he could afford to pay the premium. More frequently (for, in truth, the very thought of the London scene made him feel physically unwell), he felt an urge to drop in at the village pub, but this too he resisted. He would stick it out, alone. After a lifetime, or half a lifetime, of dissolute company, he would give solitude a fair trial.[34]

Later on the author concludes, bitterly: "People should not get together. They are more attractive in smaller groups. Collectivity corrupts. Man is a social animal, but only at great risk."

John Cotton, Boston's leading clergyman in the first half of the seventeenth century, summed up a common preference in his day: "Society in all sorts of humane affaires is better than solitariness."[35] Or, as Dálila pleaded with Samson (albeit for her own private ends):

> . . . that grounded maxim
> So rife and celebrated in the mouths
> Of wisest men; that to the public good
> Private respects must yield.[36]

Cotton had referred to a priority that arose from a strong sense of the greater threat of loneliness and isolation in a wilderness society. Americans of the nineteenth century, no longer confronted with Cotton's wilderness, tended nonetheless to lose sight of that priority but continued to seek a balance between solitary privacy and societal affairs.

The year 1850 fixes a time in which the United States was both nostalgic and futuristic, a time when Americans could look back with fond memories on simpler days, and a time when they could look forward to a more complex but progressive era of expansion (some wilderness still existed) and industrialization. Privacy before 1850, roughly, was a manifestation of geography, architecture, rural and town planning, and ethnic heritage. As is evident from Flaherty's comprehensive study, personal privacy was related more to the circumstances of one's immediate environment (namely, the home) than it was the result of any preconceived notion of a legal right, or even a moral or natural right. Privacy may have been looked upon as God-given, as in Biblical times, but it was not of dominant concern to the colonists or their leaders. Because privacy was not seriously threatened, it was taken for granted—recognized and revered by custom and circumstances. Where American society was not yet pluralistic, men less ambitious, and Jeffersonian idealism prevailed, the government that really was best was that which governed least.

Rural and agrarian still best described the nation prior to 1850, when there was little evidence of mechanical or technological development. The period accentuated the importance of the individual and minimized the role of government, if not of the larger community of citizens. If each was protected, then all were secure. And, above all, the Constitution was left to literal interpretation. As Thomas H. O'Connor writes of privacy in this historical perspective:

> A stronger spirit of isolation than ever before came into being after 1816, and most Americans enthusiastically hailed the belief that the New World had written off all its obligations to the Old World and would have no further involvement in the affairs of Europe.[31]

"Spirit of isolation" is the key phrase, which also depicted the nation's psyche, as spatial distance continued to be available to those who would heed Horace Greeley's advice, "Go west, young man, and grow up with the country," an expression he borrowed from an Indiana newspaper and reprinted in his New York *Tribune*. O'Connor posits that solitary isolation was so absolute that privacy was assured by the enormousness of the physical dimensions of the frontier.

These years were also symbolized by Yankee ingenuity, rugged individualism, personal freedom, studied isolation, effusive nationalism, and so on. Ralph Waldo Emerson's many publications included the essay "Self-Reliance," which praised the virtue of moral independence and the importance of noncomformity, and *Society and Solitude*, a volume of lectures he had presented to audiences as far west as the Mississippi. "What I must do is all that concerns *me*, not what people think," Emerson lectured.[38] But "Self-Reliance" was perhaps the most brilliant display of the writer's literary strategy and popular impact. It has been the source of frequently quoted Emersonianisms over time. "Trust thyself; every heart vibrates to that iron string." "Whoso would be a man, must be a nonconformist." "A foolish consistency is the hobgoblin of little minds, adorned by little statesmen and philosophers and divines." Emerson was to America of the nineteenth century what Benjamin Franklin, as Poor Richard, had been to an earlier generation. The individual was important and could get ahead on his own initiative, for, as Franklin exhorted, "God helps those who help themselves."

Society and Solitude, published in 1870, contains twelve essays based on slightly revised lectures, which focused on domestic life, farming, and the wise balance of solitude and sociability. Thomas Carlyle thought it Emerson's best book because of its simplicity and homely grace. As for its accuracy, however, Emerson's biographer tells of Bret Harte's visit to Concord, during which he challenged his host's statement: 'Tis wonderful how soon a piano gets into a log hut on the frontier," to be followed by a Latin grammar and other civilizing influences. Harte is reported to have said: "It is the gamblers who bring in the music to California. It is the prostitutes who bring in the New York fashions of dress there, and so throughout." Emerson replied that he spoke "from Pilgrim experience, and knew of good grounds the resistless culture that religion effects." Harte, the westerner, was not persuaded.[39]

Henry David Thoreau mused on the banks of Walden Pond, "I have, as it were, my own sun and moon and stars, and a little world all to myself."[40] From a reading of *Walden* one gets a sense of the country's transition from agrarian, and private, to an industrial, and public, society. Thoreau's cabin in the woods was only five hundred yards from the railroad between Boston and Fitchburg. The passing trains caused him to consider what was, in 1847, still a new kind of transport:

The startings and arrivals of the cars are now the epochs in the village day. They go and come with such regularity and precision, and their whistle can be heard so far, that the farmers set their clocks by them, and thus one well conducted institution regulates a whole country. Have not men improved somewhat in punctuality since the railroad was invented? Do they not talk and think faster in the depot than they did in the stage-office? To do things 'railroad fashion' is now the by-word."[41]

Walt Whitman, too, characterized America at midcentury, and the following passage gave Manifest Destiny its broadest sentiment:

It is from such materials—from the Democracy, with its manly heart and its lion strength spuring the ligatures wherewith drivellers would bind it—that we are to expect the great FUTURE of this Western World! a scope involving such unparalleled human happiness and rational freedom, to such unnumbered myriads, that the heart of a true *man* leaps with a mighty joy only to think of it![42]

Whitman, in poetry and prose, celebrated the self and sang of individual freedom against all constraints and conventions. But he also, following Franklin, Jefferson, and Emerson, expressed a strong commitment to communitarian ideals, as in his *Democratic Vistas* (1871). He wrote of the value of participation in the common life, but his greatest popular impact has been on the virtues of the private self.

This was the time of important changes in the entire political, social, economic, and cultural fabric of the country. Before the presidency of Andrew Jackson, political control had been in the hands of a comparatively small and elite group of easterners. With Jackson and later, such leadership spread across the nation, and democracy underwent a second revolution. Roads, turnpikes, canals, steamboats, and railroads connected villages to towns and cities to urban centers, blurring the previous distinctions between country and city. O'Connor, in his abbreviated history of the period, notes how technological inventiveness spurred the appearance of mechanical devices that transformed life in both the city and the farm. One has only to cite the obvious in order to appreciate the significance of the events: the reaper, the power loom, the sewing machine, the six-shooter, the rotary press, the typewriter, vulcanized rubber, the telegraph, the transatlantic cable, together with the many techniques for improving agricultural productivity and the prominence of American industry.[43]

While much of the nation remained agrarian, the entire country shifted to an industrial economy and from a rural to an urban culture. Because the changes took place in a short span of time, compared with previous shifts in the national experience, the human condition was affected profoundly in the process. How, precisely, is a matter for scholars to ponder on a scale both better focused and much grander than is intended here. But some explanation may be found in the literature of the end of the century, anticipating the first American efforts to proscribe by law invasions of solitude.

Why was privacy in need of legal protection? Were not the old Biblical traditions and colonial methods still workable? Literature, as we have seen in the brief references to Emerson, Thoreau, and Whitman, provides insight on the psychological, even metaphysical, aspects of what Americans thought about the self and, especially, the individual in relationship to others. Two more examples, these from Henry James, reveal the era's perceptions of personal autonomy amid an increasingly crowded civilization. The social environment still managed to accommodate both the individualist and the communitarian, and people were also learning to cope with a new element—the curious newspaper reporter.

Madame Merle, the pragmatic metaphysician of *The Portrait of a Lady* (1881), asks the fundamental questions: "What shall we call our 'self'? Where does it begin? Where does it end?" And she responds: "One's self—for other people—is one's expression of one's self; and one's house, one's furniture, one's garments, the books one reads, the company one keeps—these things are all expressive."[44] In her "bold analysis of the human personality," Madame Merle, "a woman of strong impulses kept in admirable order," assures Isabel, "our heroine," that the "self overflows into everything that belongs to us—and then it flows back again."[45]

Madame Merle reminds her young protégé that "every human being has his shell, and that you must take the shell into account. By the shell I mean the whole envelope of circum- stances." Madame Merle bases her philosophy on the belief that there is simply "no such thing as an isolated man or woman; we are each of us made up of a cluster of appurtenances." Thus, the notion of "self" is intrinsically tied to "other people" and their perceptions of "one's expression of one's self."[46] In this, as in other James fiction, there is a marked difference between what one as an individual may "care to be judged by" and how the individual *within society* can, and will, be judged.

The self, in Madame Merle's terms, does indeed "flow back again," but this reverse movement means, for James, that it is absolutely essential to preserve the integrity of both the self *and* the company of others. That is to say, personal privacy does not—indeed, cannot—exist or flourish in a vacuum, in spite of what Thoreau said he experienced and of Whitman's identity crisis. The second example from James is even more cogent, less philosophical, in terms of privacy invasion by the emerging popular press.

George Flack, the young American newspaperman, "saw ev- erything in its largest relations" in *The Reverberator* (1888). "There are ten thousand things to do that haven't been done, and I am going to do them," the journalist tells Francie Dosson, whom he encounters in Paris with her father and sister.

> The society-news of every quarter of the globe, furnished by the prominent members themselves (oh, *they* can be fixed—you'll see!) from day to day and from hour to hour and served up at every breakfast-table in the United States—that's what the Ameri- can people want and that's what the American people are going to have.

He goes on:

I'm going for the secrets, the *chronique intime*, as they say here; what the people want is just what isn't told, and I'm going to tell it. That's about played out, any way, the idea of sticking up a sign of 'private' and thinking you can keep the place to yourself. You can't do it—you can't keep out the light of the Press. Now what I'm going to do is to set up the biggest lamp yet made and to make it shine all over the place. We'll see who's private then![47]

Indeed, as readers of James's novelette must already have known, personal privacy in both the Biblical and colonial sense had begun to fall victim to a probing press that had become more pervasive through industrialization. Among the news readers in Boston were a couple of equally assertive lawyers whose own writing on the subject of privacy and the press would begin to stir the legal community at the turn of the century.

As for Francie Dosson, she appears impressed but not threatened by the youthful arrogance of the newspaperman Flack. She said her family sometimes saw his paper, *The Reverberator*, and that her sister, Delia, frequently read aloud from it, but, as for herself, she preferred books. "Well, it's all literature," retorted Flack. "It's all the press, the great institution of our time. Some of the finest books have come out first in the papers. It's the history of the age."[48]

Later on, true to his earlier exhortation and promise, Flack published an intimate piece that the Probert sisters, whose brother Gaston was Francie's betrothed, found "immoral" because it included "everything that's private and dreadful." To Francie's father, however, Flack's "letter from Paris appeared lively, 'chatty,' even brilliant, and so far as the personalities contained in it were concerned he wanted to know if they were not aware over here of the charges brought every day against the most prominent men in Boston."[49] Although it may only be a coincidence that the novelist would allude to the kind of press behavior that aroused Samuel D. Warren and Louis D. Brandeis, such intrusions occurred regularly in the American press, journalism of the kind not yet experienced by the aristocratic French. George Flack said he knew that the public would appreciate a column about what was going on in the *grand monde*.

When Gaston Probert threatened to end his engagement to Francie over the unfortunate episode, Mr. Dosson viewed the altercation as a failure of friendship, not the result of publication of details about the Proberts. James, who had studied law at Harvard in the 1860s, reasoned on behalf of Dosson that if "these people" had done bad things they ought to be ashamed, and "if they hadn't done them there was no need of making such a rumpus about other people knowing. If you start a paper you've got to give them what they like. If you want the people with you, you've got to be with the people." However, Gaston does chastise Francie and her American newspaper friend for the publication of "such a flood of impudence on decent, quiet people who only want to be left alone." In the end, Francie is forgiven, Flack returns home, and the Dossons and young Probert happily leave Paris for more travel, "to some place where there are no newspapers."[50] In Boston, meanwhile, a city of inquiring newspapers, the two lawyers had begun to develop their thesis, that "quiet people who only want to be left alone," as suggested by Henry James, deserved some relief in law. Before turning to that important essay, which appeared in 1890 in the *Harvard Law Review*, it is valuable to reflect briefly on the country's profile at the time.

In that year the United States Bureau of the Census officially declared the American frontier to be at an end. Foreign immigration and signs of emancipation within the nation had been at work changing the character of the population to its present pluralistic makeup. The "melting-pot," though largely unfulfilled, contained a mix of races, creeds, and nationalities, which not only altered traditional values but, more specifically, threatened old images of the individual. The historian Thomas H. O'Connor relates how the new industrial centers spawned crowded cities, overcrowded tenements, and slums whose families bore little resemblance to the farming communities of Jefferson's day. "Independence was clearly turning into interdependence by the opening of the twentieth century," O'Connor declares. "The close proximity in which members of the urban communities found themselves forced to live made individual privacy a thing of the past."[51] If not in total eclipse, at least the kind of privacy that had been equated with a more nurtured individualism of a more spacious time was clearly changing. Individualism, too, had begun to change, as is discussed in a later chapter, from the concept of public involvement to one of personal enhancement.

In this chapter, we have traced the history of privacy as simple solitude from its Biblical and classical beginnings through nineteenth-century America—the desire to withdraw from complexity. Humans have always valued solitary privacy and claim the right to seclude the self from others—to be anonymous in a crowd. But humans have also held in high regard citizen participation in society's affairs—to be visible in a crowd. Even in ideal Arcadia, privacy was never the dominant concern. But it has become so in the twentieth century, when "private" as "personal" seemed to mandate official protection.

2

Creating a Legal Right

"Privacy was a luxury undreamed of in that day, and you had little of it," Esther Forbes writes of eighteenth-century America.

> From childbirth to deathbed one's life was shockingly open to one's family, friends, relatives, neighbors, enemies, clergymen, and the curious. It was not a matter of social position. At Buckingham, the young King of England, George III, had little more privacy than a Boston artisan. Nor Louis XV at Versailles. They seemed to have had no more conception of privacy as a desirable thing than they had of electricity, and did not miss either.[1]

To a point, Forbes's portrayal is accurate (as may be discerned from Flaherty's study of colonial New England). Yet, although personal privacy may have been limited in terms of interpersonal relationships, the emphasis on the spiritual quality of life did allow for the soul and for the family to enjoy solitude. This puritanical ethos, Edward Shils believes, gave impetus to privacy in New England. Citizens may not have called it that, but they knew instinctively the value of the self. From the inner to the outer self, what has been called "growth of individuality," was the transformation that enhanced the belief, in Shils's words, "that one's actions and their history 'belonged' to the self which generated them and were to be shared only with those with whom one wished to share them."[2]

Ironically, what are now seen as major influences *against* privacy—urbanization, industrialization, literacy, and education—also have made the maintenance of privacy easier. Shils's perception is helpful: "People did not cease to be interested in their neighbors; but they had to contend with the increased resistance of their neighbors to being known and with increased difficulties in knowing about them." Improved literacy, increased education, and greater involvement in politics meant that the citizen's radius of interest and attention extended beyond the neighborhood, the town, and, as time went on, the country. Government, too, paid more attention to the rights of citizens, with privacy eventually falling within constitutional boundaries. As Shils notes: "The respect for private property that the state was concerned to enforce helped to stiffen a general regard for privacy, and the ethos of economic individualism worked in the same direction." By the end of the nineteenth century, property, privacy, and individualism had closely related or interdependent meanings.[3]

Printing, telegraphy, and photography were technologies that at the turn of the century were seen as both marvels of the age and instruments of disruptive change. The fears they generated were in many ways the same as those in our own time of radio and television, and electronics in general. Individual privacy then, as now, was perceived to be threatened by these mechanical forces. "The rapid advances of technology," O'Connor writes, "made it more possible than ever before in history not only to satisfy the demands of public curiosity, but to excite and inflate that curiosity to even greater dimensions."[4]

Constitutional jurisprudence failed to provide needed relief from aggressive journalism. Jurists simply refused, and with sound historical reason, to depart, as O'Connor points out, from a rigid and literal interpretation of the Bill of Rights. Privacy as a right did not exist in British common law, nor was privacy, as such, mentioned in the U.S. Constitution, but the Bill of Rights assured such aspects of privacy as religious belief and practice.

By the end of the nineteenth century the American press had

grown to the point where everyone who could read (or desired to be read to) had access to newspapers on a daily basis. The emergence of cheap gazettes, journals, tribunes, and heralds by midcentury created a small but significant revolution in social life, not to mention that the press itself was an economic and political force. During a twenty-year period near the end of the century the number of newspapers doubled to a total of 921. More than six hundred were established between 1880 and 1889 alone. Circulation increased by nearly 1,100 percent between 1850 and 1890. At midcentury, 758,000 newspapers circulated each day; by 1870, the figure was 2,607,000; and by 1890, the circulation of newspapers in the United States had reached 8,387,000 copies daily. Curiosity had always been a human trait, but the newspapers stimulated popular interest in private affairs far beyond that of any earlier time.

The industrial revolution had stimulated mass-production methods in the press, as it had in other businesses, and the consequence was a new kind of populist journalism that featured sensational, colorful, and, in the end, more interesting news. Much of the content was of a personal nature, depicting the lives of the upper crust as well as those whose names appeared on police records and domestic court dockets. Some citizens viewed this development in American journalism as a menace to society in general and to the individual in particular. So strong was this view that newspapers were openly criticized for their frequent waywardness.[5]

A writer in the *Nation* magazine in 1873, for example, complained that the interviewing technique used by journalists was an affront to men of stature. Through interviews of leading public figures, "newspaper correspondents are driving the public . . . into wondering that a sage can be such an ass."[6] President Cleveland expressed dislike of the way the press treated him on occasion, especially when some journalists followed the President and his bride on their honeymoon trip in 1886. A few months later, Cleveland, in an address at Harvard, said journalists were purveyors of "silly, mean, and cowardly lies . . .

and in ghoulish glee desecrate every sacred relation of private life."[7] E. L. Godkin wrote in *Scribner's Magazine* in July 1890 that the chief enemy of privacy in modern life was the curiosity shown by some people about the affairs of other people. "In all this," Godkin wrote, "the advent of the newspaper, or rather of a particular class of newspaper, had made a great change. It has converted curiosity into what economists call an effectual demand, and gossip into a marketable commodity."[8]

The right of privacy in American common law has its roots in the famous Warren and Brandeis article of 1890, which propounded the concept as independent from the rights of property, contract, and trust.[9] Unlike these rights, long recognized in common law, the right of privacy, as envisioned by Samuel D. Warren and Louis D. Brandeis, sought to protect the plaintiff's right "to be let alone." The essential injury, they said, was to one's feelings of consequent mental anguish. With their essay, the legal community started to develop an argument based on the individual's right not to have his thoughts, statements, or emotions made public without his consent. Roscoe Pound, then dean of the Harvard Law School, later commented that Warren and Brandeis had done "nothing less than add a chapter to our law."

Warren and Brandeis argued that, whereas some aspects of privacy may involve ownership or possession, privacy to them meant "inviolate personality," and what they sought was a way to protect the personality itself. They did not reject the property aspects of privacy, but rather pleaded for a distinct right. Theirs was a legal remedy for the anguish caused an individual where there had been no injury to property or contract rights. Warren and Brandeis wanted to extend the scope of jurisprudence to protect not the reputation—already covered by libel law—but the feelings of individuals who had been subjected to some form of intrusion.

Their approach is believed to have grown out of their "abhorrence of the invasions of social privacy" by the newspapers, as Brandeis said in a note to his close friend and former partner,

Warren. Brandeis's official biographer attributes the idea for the article to Warren's annoyance over a particular report in "lurid detail" of his family's Back Bay social activities. But there is reason to believe that Godkin's *Scribner's* piece, as well as the famous editor's reputation among intellectuals, contributed to their thinking. Of Godkin, William James had written: "To my generation his was certainly the towering influence in all thought concerning public affairs, and indirectly his influence has assuredly been more pervasive than that of any other writer of the generation."[10] Indeed, Warren and Brandeis cited Godkin, "an able writer," in their privacy article. They also credited Judge Thomas Cooley, who, in his famous treatise on torts, published in 1879, spoke of the right "to be let alone" as a matter of personal security. "The right to one's person may be said to be a right of complete immunity," Judge Cooley said.[11]

Godkin, who had founded the *Nation* in 1865 and then suc- ceeded William Cullen Bryant as editor of the New York *Evening Post* in 1881, was known widely as a civic reformer. In his *Scribner's* essay on privacy, he traced the history of man's desire for privacy from the days of communal life in wigwams to the civilization of his own time. He said both rich and poor desired private homes, which demonstrated "the ambition of nearly all civilized men and women . . . [to decide] for themselves how much or how little publicity should surround their daily lives."

When Anglo-American law recognized a man's house as "his place of repose," this was

> but the outward and visible sign of the law's respect for his per- sonality as an individual, for that kingdom of the mind, that inner world of personal thought and feeling in which every man passes some time, and in which every man who is worth much to himself or others, passes a great deal of the time.

Although the importance attached to privacy "varies in indi- viduals," Godkin said, those who do seek and need privacy "are the element in society which most contributes to its moral and intellectual growth."[12]

Godkin saw curiosity as the chief enemy of privacy. Oral gossip was not the culprit, for that was limited to a few people and could be ignored, but the development of mass-circulation, sensation-seeking newspapers had transformed curiosity into a marketable commodity. Libel law did not go far enough, for Godkin doubted whether juries would award damages for "mere wounds to (a man's) feelings or his taste." Although he saw public censure of press misconduct as a possible solution, he believed that neither law nor opinion would be inclined to protect the sensitive man's claim to privacy. The trick, according to Godkin, was to achieve control of invasions of privacy without either the unlikely support of public opinion or suppression of the press. That task was taken up by Warren and Brandeis, Harvard law graduates with access to their alma mater's prestigious journal.[13]

"That the individual shall have full protection in person and in property is a principle as old as the common law; but it has been found necessary from time to time to define anew the exact nature and extent of such protection." Warren and Brandeis described the common law as having evolved steadily to protect man from physical interference with life and property. This "right to life," they noted, served only to protect the individual against trespass and battery. But, as Judge Cooley had asserted, "battery involves many elements of injury not always present in breaches of duty. There is very likely a shock to the nerves, and the peace and quiet of the individual is disturbed for a period of greater or less duration."[14] Warren and Brandeis traced these early actions to such concepts as nuisance, slander and libel, and intangible property, the "products and processes of the mind." However, new inventions and business methods required that a next step be taken to protect the right "to be let alone."

The authors then detailed their criticism of the press.

Instantaneous photographs and newspaper enterprise have invaded the sacred precincts of private and domestic life; and numerous mechanical devices threaten to make good the prediction that 'what is whispered in the closet shall be proclaimed from the

house-tops.' The press is overstepping in every direction the obvious bounds of propriety and of decency. Gossip is no longer the
resource of the idle and of the vicious, but has become a trade,
which is pursued with industry as well as effrontery. To satisfy a
prurient taste the details of sexual relations are spread broadcast in
the columns of the daily papers. To occupy the indolent, column
upon column is filled with idle gossip, which can only be procured
by intrusion upon the domestic circle.

Next, Warren and Brandeis explained their underlying rationale, a commentary of sorts on American society at the time.
"The intensity and complexity of life, attendant upon advancing
civilization," they said, "have rendered necessary some retreat
from the world, and man, under the refining influence of culture,
has become more sensitive to publicity, so that solitude and
privacy have become more essential to the individual; but modern enterprise and invention have, through invasions upon his
privacy, subjected him to mental pain and distress, far greater
than could be inflicted by mere bodily injury." The spreading of
gossip both belittled and perverted the relative importance of
things, "dwarfing the thoughts and aspirations of a people. . . .
Easy of comprehension, appealing to the weak side of human
nature which is never wholly cast down by the misfortunes and
frailties of our neighbors. . . . Triviality destroys at once
robustness of thought and delicacy of feeling. No enthusiasm can
flourish, no generous impulse can survive under its blighting
influence."

Some scholars have challenged the legal implications of the
Warren-Brandeis thesis, and still others, most notably Don R.
Pember, a communication law professor, have questioned the
authors' heavy-handed portrayal of the "evil" for which they
believed a remedy was sorely needed.[15] The legalisms are examined later in this chapter, but it is important here to point out a
basic inconsistency that emerges in the Warren-Brandeis thesis.
On the one hand, their reasonable philosophy supports unques-
tionably an ideal civilized and humane society's interest in limiting public discussion of private matters. On the other, there is
the need in an open society for as much information as possible

to circulate without penalty, and sometimes the information is of a private nature. It is basically, as Professor Diane L. Zimmerman asserts correctly, the "challenge of harmonizing privacy with free speech."[16] It is also, as Sir Patrick Devlin and H. L. A. Hart were to debate later, a matter of whether society has the right to enforce a morality on the ground that a shared norm is essential to society's existence.

In advancing their thesis, Warren and Brandeis made no effort to hide the fact that they were most interested in the moral, or spiritual, not the material, side of law. "Our law recognizes no principle upon which compensation can be granted for mere injury to the feelings." They found such a principle in the doctrine of common law copyright, which gives the author or artist exclusive ownership. But, they said, the statutory right is of no value *unless* there is publication and the common law right is lost *as soon as* there is publication. Thus, Warren and Brandeis summarized: "The principle which protects personal productions . . . against publication in any form, is in reality not the principle of private property, but that of an inviolate personality." They deduced, therefore, that "a general right to privacy for thoughts, emotions, and sensations . . . should receive the same protection, whether expressed in writing, or in conduct, in conversation, in attitudes, or in facial expression." The right to privacy, as a part of Judge Cooley's more general right to the immunity of the person, is the right to one's personality.

Milton R. Konvitz has drawn a more than subtle distinction. Cooley's phrase, "the right to be let alone," is preferable to the Warren-Brandeis phrase, "the right to privacy." Konvitz suggests that the latter is at once more restrictive and more general, for it suggests what has been withdrawn from public view—the marital bedroom or a respectable married woman's past immoral life. It suggests secrecy and darkness, elements of the private life that are detrimental to public welfare. Privacy, Konvitz notes, may also be essential to acts performed in public view, such as membership in an organization or worshiping in a

church or synagogue. A person may be asserting his or her right
of "privacy" when they dress in an unorthodox way or when
they "loaf" in a public park. "A person may claim the right to be
let alone when he acts publicly or when he acts privately," Konvitz points out. Judge Cooley's right implies the kind of space a
person may carry anywhere, into the bedroom or into the
street.[17]

Perhaps Warren and Brandeis anticipated the kind of problems Konvitz discerns, for they placed several limitations on
their new right. First, the right of privacy did not prohibit the
publication of any matter of public or general interest. "Public
interest" is central to the philosophy of the First Amendment,
part of the political freedom James Madison championed when
he engineered ratification of the Constitution and the Bill of
Rights. But the term has always escaped precise definition and
easy administration. "The design of the law must be to protect
those persons with whose affairs the community has no legitimate concern," Warren and Brandeis said. "It is the unwarranted invasion of privacy which is reprehended. The general
object . . . is to protect the privacy of private life." They said the
right may be withdrawn, however, if a man's life ceases to be
private. With this concession to freedom of expression, Warren
and Brandeis created a Pandora's Box for constitutional scholars
and jurists. They left for future generations the impossible task
of deciding "legitimate concern" and "unwarranted invasion."

Second, Warren and Brandeis insisted that their right did not
prohibit the communication of any matter, "though in its nature
private," under circumstances that would render it a privileged
communication as defined by the law of slander and libel. This
would include statements made at public meetings or in courts,
the publication of which clearly serve a public purpose.

Third, their law of privacy did not grant redress for any wrong
suffered by oral "publication" of private matter. "The injury
resulting from such oral communications would ordinarily be so
trifling that the law might well, in the interest of free speech,
disregard it altogether." Gossip, in other words, could be toler-

ated so long as a person's "peace and comfort were . . . but slightly affected by it," as E. L. Godkin had allowed.

Fourth, the right ceased when the individual published the facts himself, or consented to their publication. Truth, long a defense in libel actions, did not afford a defense in a privacy suit, the authors said. "It is not for injury to the individual's character that redress or prevention is sought, but for injury to the right of privacy." Nor, finally, was the absence of malice an acceptable defense. Warren and Brandeis said that their right was "equally complete and equally injurious," regardless of the motives.

As for legal remedies, the authors suggested two: financial compensation for any injury to the plaintiff's feelings; and injunctive relief designed to stop an invasion but with no compensation for damage already incurred. Although they did not spell out the kinds of cases that might require an injunction, one example is that of a newspaper's use of a photograph in an advertisement that causes damage that cannot be corrected after publication. The authors concluded their forceful argument:

> The common law has always recognized a man's house as his castle, impregnable, often, even to its own officers engaged in the execution of its commands. Shall the courts thus close the front entrance to constituted authority, and open wide the back door to idle or prurient curiosity?

Although Warren and Brandeis sought legal remedies for a particular kind of intrusion, American courts and legislatures had for some time recognized the home, confidential communications, and public records as private domains. "A man's house is his castle" was both a popular proverb and a legal maxim in the nineteenth century. Sir William Blackstone, whose *Commentaries on the Law of England*, published between 1765 and 1769, influenced American common law, said that the law had a special regard for the "immunity of a man's house." Blackstone advocated criminal punishment for public nuisances and eavesdroppers, and he established the rule that government officials could not break down doors for the execution of any civil process. Thus, American courts administered criminal penalties and civil remedies to safeguard the "sanctity and inviolability of one's house," the householder's right to "quiet and peaceable possession," and the dwellinghouse as "the place of family repose." The law of trespass and the constitutional prohibition of unreasonable search and seizure—the Fourth Amendment— were interpreted as protection against official and unofficial intrusion.[18]

In addition, public opinion regarded the "sanctity of the mails" as absolute in the same way it esteemed the inviolability of the home. A Post Office agent explained in 1855 that "the laws of the land are intended not only to preserve the person and material property of every citizen sacred from intrusion, but to secure the privacy of his thoughts, so far as he sees fit to withhold them from others." A letter was considered the property of the sender while en route, but the unauthorized opening and reading of its contents was neither theft nor damage to property. Congress nevertheless made interception intended "to pry into another's business or secrets" a criminal offense in 1825, a statute believed to have been drafted chiefly by Daniel Webster. Once the letter was delivered, this "sacred epistolary communion" was protected by law in some states.[19]

The telegraph, the "distance-killing gadget" in full use by the 1860s, presented another kind of problem because messages were necessarily read by the operators who sent and received them.[20] Wiretapping, a practice learned by military telegraphers during the Civil War, was deemed a crime in some states, and elsewhere it was viewed as illegal interference with telegraph company property. Such laws were intended "to prevent the betrayal of private affairs . . . for the promotion of private gain or the gratification of idle gossip."[21]

Other private conversations and sensitive items of personal information were protected from courtroom disclosure and in official records of legal proceedings. The common law rule of evidence that excluded spousal testimony and confidential communications between husband and wife, rooted in old notions of the couple's single legal identity, acquired a new justification in the nineteenth century. Several cases noted the value in preserving "the sacred privacy of domestic life." Courts went far to avoid "embarrassing questions" that would "tear . . . away the veil, which hides from public gaze the sacred confidences which subsist between husband and wife." The same common law privilege applied to attorney-client communications and to disclosures of information between patient and doctor and penitent and priest.[22]

"Mind your own business" was another nineteenth-century proverb that acquired judicial attention as more and more personal business found its way into public records and into the columns of enterprising newspapers. Common law doctrines often failed to protect unwanted disclosure, but the courts and legislatures restricted access to information that individuals were required to supply the government and designed broad civil and criminal remedies against the press.

In the first U.S. census in 1790, the government required little more than the enumeration of persons, slave and free, but even then there was opposition on privacy grounds. Citizens continued to object as each census asked for more and more personal information, resulting in instructions to census takers in 1840 that individual returns be treated as confidential. Congressman James A. Garfield, the chief advocate of a statutory penalty for disclosure of census data, feared that "the citizen is not adequately protected from the danger, or rather the apprehension, that his private affairs, the secrets of his family and his business, will be disclosed to his neighbors," or, at the central office, be "made the quarry of bookmakers and pamphleteers." A penalty for such disclosure was enacted in 1889.[23]

Outrageous or offensive publications of private matter could be prosecuted under libel laws, but defamation suits further publicized unwanted disclosure and focused trials on the issue of truth or falsity of the damaging material. As Warren and Brandeis observed correctly, libel doctrines did not reach new invasions made possible by advances in photography in the late 1880s, such as the taking of "candid" pictures and their subsequent publication without the subject's consent.

Meanwhile, the *Harvard Law Review* article was cited within a year of its publication in a suit brought by an eminent British physician whose name had been used without permission in advertising a remedy for catarrh that he purported to endorse. It was again cited two years later by a New York court in an actor's suit against a newspaper publisher for using a picture of the plaintiff and another actor and inviting readers to vote who was the more popular. Both cases involved professional reputations, not privacy per se, but the New York court quoted Judge Cooley's precept, that the plaintiff had a "right to be let alone."[24]

The first reported case in which the right of privacy was expressly recognized was in 1892, when a New York judge granted an injunction against "the making and public exhibition of a statue of a deceased person . . . where it is not shown that she was a public character." In distinguishing between private and public persons, the judge said:

> The moment one voluntarily places himself before the public, either in accepting public office, or in becoming a candidate for office, or as an artist or literary man, he surrenders his right to privacy, and obviously cannot complain of any fair or reasonable description or portraiture of himself.[25]

The ruling was novel for the time, for it extended the right of privacy beyond that of earlier cases and even beyond that envisioned by Warren and Brandeis.

They had argued that existing doctrines provided the basis for an independent right of privacy to be recognized, especially those found in the law of intellectual property, or copyright. Warren and Brandeis thought of privacy as some form of property. They also insisted that the right becomes more necessary as technology advances, symbolized in their time by the proliferation of newspapers. But, when their specific thesis was applied ten years later, it failed.

Right of privacy was the issue in *Roberson v. Rochester Folding*

Box Co. in 1902,[26] which involved the unauthorized use of a girl's portrait on 25,000 posters. Although the picture had been used without her consent, the four-to-three New York Court of Appeals decision denied her an injunction on the ground that no right yet known to common law had been infringed. The court said that none of the cases cited by Warren and Brandeis applied to the plaintiff solely on the basis of hurt feelings. However, Judge John C. Gray, in dissent, linked personality to property within the right of privacy created by Warren and Brandeis. "I think that this plaintiff has the same property in the right to be protected against the use of her face for defendant's commercial purposes as she would have if they were publishing her literary compositions."[27] Thus, the right of privacy, as promulgated by the two Boston lawyers, carried at least an implied right of property.

Subsequent cases that recognized the new concept also recognized the basic common law right of property in one's name and likeness. Most decisions held that a plaintiff's name and likeness could not be appropriated in a purely commercial venture for the profit and gain of the appropriator. A majority of the cases alleged humiliation and mental anguish, so the courts usually applied the independent right of privacy thesis under which property ownership is not a factor.

The Roberson decision "excited as much amazement among lawyers and jurists as among the promiscuous lay public," as the *New York Times* editorialized, but it was also defended. Judge Denis O'Brien, writing in the *Columbia Law Review* (November 1902), explained his interpretation of the ruling:

> The right of privacy in such cases, if it exists at all, is something that can not be regulated by law. The rules for the regulation of human conduct with respect to the courtesies and proprieties of life and that enjoin that delicate regard for the feelings and sensibilities of others are not to be found in statutes or judicial decisions.

Judge O'Brien went on to explain that feelings and sensibilities are probably violated by pictures or photographs to sell newspapers (or flour), but that the "practical question is whether such devices for stimulating business and making money can be regulated by law without doing more harm than good." Courts are duty-bound not to make new laws, O'Brien said, but to enforce those that already exist. Whenever new laws become necessary to suppress some evil, "resort must be had to the legislature."

But even when that process is the more appropriate way, it is wiser to permit the police to act against oglers and starers than to attempt to regulate their indecent conduct by statute. "If the use of this young woman's picture was a legal injury at all, it was either an injury to her person [an assault and battery charge] or to her character [a defamation charge]." The judge questioned how it was possible to harm a woman's feelings by depicting her beauty. "A woman's beauty, next to her virtues, is her earthly crown, but it would be a degradation to hedge it about by rules and principles applicable to property in lands or chattels." O'Brien also anticipated an increase in litigation on such matters. "When a court embarks on the business of making law to suit a particular case it is difficult to stop, as one decision generally furnishes an argument for another."[28] That is precisely what happened. Professor William Prosser was to write some years after the Warren and Brandeis treatise: "Although there was at first some hesitation, a host of other legal writers have taken up the theme, and no other tort has received such an outpouring of comment in advocacy of its bare existence."[29]

The majority view in *Roberson* was to be completely wiped out three years later by what became the leading privacy case, *Pavesich v. New England Life Insurance Co.* (1905).[30] The decision marked the first recognition of the right of privacy by a state court where no statute existed. The insurance company had published an advertisement in the *Atlanta Constitution* containing photographs of two men, one of whom was Paolo Pavesich, an artist, depicted as a person who had bought sufficient life insurance. The other photo, of an ill-dressed and sickly-looking man, illustrated a person who had no insurance coverage. Pavesich's picture carried a testimonial in praise of the company. The artist sued, and the Georgia Supreme Court held that "the form and features of the plaintiff are his own."

What made the case doubly significant was Pavesich's public position as a noted artist. "It is not necessary to hold," the court said, "that the mere fact that a man has become what is called a public character, either by aspiring to a public office, or by exercising a profession which places him before the public, gives to every one the right to print and circulate his picture."[31] In its unanimous decision, the Georgia court established a strong precedent for one aspect of the privacy concept, the unauthorized use of an individual's picture.

The *Pavesich* court, in viewing the right of privacy as having its foundation in the "instincts of nature," underscored the spiritual quality of individualism and also tried to distinguish between "matters private" and "matters public." Both have the force of law behind them, but the greater of the two is the right in matters purely private, for that right is derived from natural law. "Each individual as instinctively resents any encroachment by the public upon his rights which are of a private nature as he does the withdrawal of those of his rights which are of a public nature." The court had said that liberties derived from natural law, such as the right of privacy, should also be recognized by legislated, or man-made, law. Privacy, as a basic liberty, is an *endowed* human faculty, "subject only to such restraints as are necessary for the common welfare." Conversely, human faculties associated with public matters, or the common welfare, are *bestowed* by the government. Such distinctions, between Creator-endowed liberties and government-bestowed rights, have dogged attempts to legalize personal privacy from the beginning.

Pavesich is as much a philosophical plea for the life of seclusion, the private life, as it is a legal ruling, although it probably reflected general public concern for privacy in the early part of the twentieth century. The ruling actually went beyond what Warren and Brandeis had urged. Where law normally requires as much specificity as possible, the Georgia ruling had so much general bearing as to render application of its concept of privacy impossible. As Diane Zimmerman noted in her recent analysis of the failure to apply successfully the Warren-Brandeis tort: "Some human problems are impervious to legal solution because they involve social ideals that do not readily translate into intelligible legal theory." Some elude legal resolution because, in her words, "we cannot clearly identify and balance the relevant social and moral values." In this, Zimmerman echoes Judge O'Brien, who, in reaction to the earlier *Roberson* decision, said privacy could not be regulated by law. *Pavesich,* as does the entire Warren-Brandeis thesis, raises more questions than are necessary if protection is to be granted those aspects of private life that are worthy of legal help. Broad-sweeping court rulings tend to neutralize effective legal protection.[32]

Criticism of the *Roberson* decision was credited with passage of the New York "Civil Rights" statute in 1903. Since then nearly forty jurisdictions, mostly state courts, appear to recognize the Warren and Brandeis public-disclosure-of-true-facts suits. However, the New York law is limited to misappropriation invasions, where the offense is one of unjust economic enrichment and property interest is being protected. The statute says, in part:

> A person, firm, or corporation that uses for advertising purposes, or for the purpose of trade, the name, portrait or picture of any living person without having first obtained the written consent of such person or, if a minor, of his or her parent or guardian, is guilty of a misdemeanor.[33]

Following passage of the law, actions occurred in New York state that were designed to carry the interpretation of the right of privacy into nonadvertising parts of a newspaper. But the courts have said that such rulings did not necessarily apply to the dissemination of general information. "The publication of matters of public interest in newspapers or newsreels is not a trade purpose within the meaning and purview of this statute."[34] Warren and Brandeis had recommended the same limitation—their reconciliation with First Amendment theory and practice.

New York's prototype statute purported to secure in law the right of privacy, but, ironically, it neither makes intrusion into private affairs the essence of the wrong, nor does it prohibit the interruption of an individual's seclusion or solitude. And, since these were the primary concerns of Warren and Brandeis, the application of the private-facts tort continues to confound jurists. Instead, the New York law limits invasion to commercial exploitation and allows for the publication of news and information of public interest. In the final analysis, perhaps this is the extent to which personal privacy can be protected legally, if not also realistically, considering the historic reach of the First Amendment. As Professor A. E. Dick Howard recently described James Madison's chief concern with *political* freedom in writing the Constitution:

> He and his contemporaries would hardly have argued that the First Amendment could be used to limit the reach of obscenity laws, or that the right to privacy guaranteed access to contraceptives. Today's individual liberty cases owe more to John Stuart Mill than to Madison.[35]

The case of William James Sidis, which is mentioned later in the chapter dealing solely with mass media invasion of privacy, is especially illustrative of the dilemma created by efforts to legalize privacy. Or, as Frederick Davis assessed the problem, "When protected by law these interests become rights." In their broad claim for law, Warren and Brandeis referred to interests that today are better protected by various property laws, or by libel laws that cover the entity called "reputation." Davis sees the difference between "reputation" and "privacy" to be so slight as to make a separate privacy tort appear redundant, because pride is the prime motivator in both kinds of suits.[36] The Sidis case is illustrative.

Sidis had been a nationally famous child prodigy in 1910 and had lectured at Harvard, which graduated him with honors and

considerable fanfare at the age of sixteen. Following predictions
that young Sidis would become a world-renowned mathemati-
cian, he soon vanished from public view—that is, until 1937
when a *New Yorker* magazine article reported on his life as a
clerk and a collector of streetcar transfers. He told the reporter
that the world just would not let him alone. "The very sight
of a mathematical formula makes me physically ill," he said.
"All I want to do is run an adding machine, but they won't let
me."[37]

Distressed over the published article, Sidis sued the magazine, claiming a cause of action under both the common law
right of privacy and the New York statutory prohibitions. Judge
Charles E. Clark of the Second United States Circuit Court of
Appeals held that there was a legitimate public interest in the
present circumstances of a person who had once, whether wit-
tingly or unwittingly, commanded widespread attention and
who had shown such great promise: "His subsequent history,
containing as it did the answer to the question of whether or not
he had fulfilled his early promise, was still a matter of public
concern." Judge Clark also referred to the state law, which he
found restrictive enough to deny grounds for recovery. But even
had Sidis claimed a more direct property interest in his name
and personality, based perhaps on a forthcoming autobiography, the ruling would have been governed by the First Amendment's protection of such obvious noncommercial speech.

Sidis presented the classic encounter, involving press freedom, on the one hand, and just the kind of press sensationalism
that Warren and Brandeis were determined to stop. But the
subsequent law of privacy could not cope with the concepts of
"peace of mind" and "freedom from mental aggravation,"
which Sidis claimed. The case simply lacked sufficient specificity, in other words, to overcome the stronger public interest as
implied in the First Amendment. Privacy, at best a correlative
right, is more easily calculated as, in Davis's view, "an interest
or condition which derives from and is automatically secured by
the protection of more cognizable rights." It could be argued

that the "right" to privacy is more a sociological notion, even a theological construct, than it is a viable legal concept. It has about as much utility, as Jeremy Bentham's musings on such "nonsense" are discussed in a later chapter, as the "pursuit of happiness" or "natural justice" or a myriad of other democratic ideals. "Invasion of privacy is, in reality, a complex of more fundamental wrongs," Frederick Davis wrote more than twenty-five years ago. "Indeed, one can logically argue that the concept . . . was never required in the first place, and that its whole history is an illustration of how well-meaning but impatient academicians can upset the normal development of the law by pushing it too hard."[38]

Rights, while familiar to everyone, as in "property rights" or "the right to privacy," seldom hold always the same meaning in all contexts. There are also "civil rights," "the right to life," "birth rights," and "the right to know." Historically, Alan R. White tells us, it was customary in legal circles to assume that the word "right" applied to obvious things, corporeal entities such as property or land. In the nineteenth and the early twentieth century, philosophers believed the word denoted some mystical or supernatural power or bond. Bentham, among those who challenged the concept of natural rights, said "right" denotes a fictitious object (entity). White, in our own time, says that *a right* may be understood in association with other notions, or objects, such as obligation, duty, liberty, privilege, power, and claim.[39]

At the strictly conceptual level, rights may be identified as either "active" or "passive," a right to do something (to worship as one pleases) or a right to have something done (to be cared for in old age). Active rights, as perceived by White, are in the hands of the actor or agent, and passive rights are controlled by the recipient. But there can also be a right to be in a certain state (to be free or happy), a right to feel something (to be annoyed or proud), and a right to take a certain attitude or hold a certain opinion (to assumptions, expectations, or hopes).

In all categories, in order for a right to be a right the determining factor is control or power, either in the hands of the actor or the recipient—it is the product of and a participant in an interactive process. It is inconceivable, for example, that one can have a right to succeed or fail, to ache or itch, to imagine or dream. None is controlled by either the actor or the recipient. The so-called "right to privacy" is a most interesting case in point. The notion of privacy has been based for nearly a century on the "right to be let alone." This "right" embraces all of these conceptual rights, although, as has been argued, the right *to* or *of* privacy (active as well as passive) is a principle whose obstacles may well be insurmountable in terms of sound legal policy.

In the real world, rights are distinguished as either moral or legal, religious or political, statutory or constitutional, customary or conventional, epistemological or logical, and so on. A person or group may have a moral, but not necessarily a legal, right (perhaps even an obligation or duty) to oppose a certain measure. He or she may have a constitutional right to vote, or a customary right to a certain pew in church. As White notes:

> Many rights, which have their origin in the particular circumstances of the case, such as a right to criticize or to complain, to smile or to express surprise, to feel disappointed or pleased, to hope or to expect are not usually or necessarily classified as either moral or logical, constitutional or conventional.[40]

The right to privacy operates within both a moral, or conventional, and a legal, or logical, framework. Privacy, in concept and by tradition, is a basic human right; as a legal entity based in social reality, however, it is ambiguous at best. The concept is important, yet it has acquired different, at times contradictory, meanings.

Privacy, like patriotism, tends to be a general right that is difficult to accommodate except in specific circumstances or situations. If human rights—such as the right to be let alone—are to enjoy the force of law, it seems desirable for that right to be spelled out carefully. As shown in Chapter 3, the basic question with personal privacy as a legal right is its pervasiveness, its lack of specificity.

The whole idea of a right, as contrasted with what is right, is a rather modern phenomenon; it was unknown to the Greeks and possibly the Romans. White ascribes its late arrival as partly due to the delayed, but now pronounced, prominence given to the individual:

> This emphasis on the individual and what he is justified in having may also explain the current fashion for translating judgements about the right treatment of babies, foetuses, animals, and even inanimate nature into judgements about their rights.

It is also true that the concept of duty, which dates to classical times as well, emphasizes the individual. But, as White points out, it does so more by stressing what is expected or demanded of him than what he can himself expect or demand.[41]

Richard E. Flathman identifies three basic historical reasons for the prominence of rights in the moral and political life of contemporary society. First, the great emphasis on rights— human civil rights especially—is part of the American tendency to convert moral and political questions into legal ones. Second, rights are a basic and dominant concept in Western thought in general, and, as a result, rights automatically impart rightness or goodness in human behavior. The third reason stems from liberalism's rationale for the rise of capitalism, the capitalist system, and, in Flathman's words, "the entrepreneurial activities of the bourgeoisie."[42]

Hobbes, then Locke, and eventually Rousseau gave credence to the individual and the individual's role in the larger community. Flathman traces the idea of rights, particularly those of the individual, to the philosophy of "liberal individualism" that developed in France, England, and colonial American in the seventeenth and eighteenth centuries. Every individual had what Hobbes called the "right of nature," the right to pursue the satisfaction of his desires with little regard for the consequences for other persons. In the real world of human affairs, however, it is difficult to apply this theory of the individual as something apart from historical, social, or cultural contexts. Yet, as Flathman notes, rights became increasingly prominent as the liberal

sentiment that recommended them was adopted by the populace of Western societies.

According to Marxist theory, individual rights served to free
the bourgeoisie from the old feudal restrictions, but then were
used to protect the exploitative property-owning class against
the threats and demands of the new proletariat class. Flathman's summary is helpful:

> The continued, indeed increasing, prominence of *rights* is due in
> part to the surprising tenacity and even growth of capitalism and
> the bourgeoisie. It is also due in part to the associated failure of the
> proletariat to realize that *rights* is a concept tied to the capitalist
> system; that rights are granted to members of the working class
> only in name, not in fact, and only to buy off any inclinations to
> revolution that develop among them.[43]

Thus, the prominence of rights in today's Western societies
may be explained as: 1) the emergence of "legal" rights as solutions to moral and political problems; 2) the rise of "natural"
rights as essential to the individual's growth and fulfillment in
society; and 3) the emphasis in our own time on "economic"
rights, which can be exploitative or anti-exploitative and di-
visive. Clearly, the last reason has not yet met with as much
favorable acceptance as have the others, yet there is little doubt
that wealth, or its absence, is a major factor in the exercise, if not
enjoyment, of individual rights. In the case of the right to privacy, the initial effort to legalize such a right came out of intrusion into upper-class society by the middle-class representatives
of the press.

Personal privacy encountered some early resistance in evolving
as common and legislated law, but it acquired more acceptance
in constitutional law. The Fourth Amendment's strictures
against "unreasonable search and seizure" was thought to be a
question of privacy when the U.S. Supreme Court decided a
case in 1928 involving the use of wiretapping equipment by the
police as a way of obtaining evidence. Chief Justice William
Howard Taft, writing for the majority, ruled that because there

had been no actual entry into the houses, the search and seizure amendment did not apply.

Louis Brandeis, by then an associate justice on the Supreme Court, joined Oliver Wendell Holmes in a famous dissent. To Holmes wiretapping was "dirty business" because of the shabby role played by the government. "I think it less evil that some criminal should escape than that the government should play an ignoble part," he said. Brandeis, predictably, was disturbed by the effect of this *Olmstead* decision upon "the right to be let alone—the most comprehensive of rights and the right most valued by civilized men." This itself is a strong reaffirmation of the views he and Samuel Warren had expressed in their 1890 article, but in *Olmstead* it was the government, not the press, whose behavior he scorned.

> To protect that right, every unjustifiable intrusion by the Government upon the privacy of the individual, whatever the means employed, must be deemed a violation of the Fourth Amendment. And the use, as evidence in a criminal proceeding, of facts ascertained by such intrusion must be deemed a violation of the Fifth.[44]

With this, personal privacy became a matter of constitutional law; and the government had been identified by the court as a potential privacy invader. Subsequent wiretapping cases have been dealt with by the high court, but in most, whether decided for or against the government, the decisions have hinged on the *technique* involved and not on the *principle* of such official surveillance. Privacy rulings, then, have provided no clear theory on which decisions could be based. Because legislation lagged behind society's increasing concern over personal privacy, the courts were left with the responsibility of identifying a legal umbrella for invasion cases.

At the time Warren and Brandeis broke with legal tradition, the country was having to deal with a number of other upheavals in intellectual traditions. Historical research had turned to science as its model. Other changes were wrought by Darwinian evolution and the emergence of psychology, sociology, and economics as fields of systematic study and influence. Also, the new liberal ideology, which was antiformalist, evolutionary, and relativist, gave renewed emphasis to the relationship of the individual to the rest of society. Such figures as Thorstein Veblen, John Dewey, W. G. Sumner, James Harvey Robinson, Charles Beard, and Oliver Wendell Holmes were at work re-defining economics, philosophy, history, and jurisprudence in the light of contemporary social and physical circumstances of the time.

Holmes, from his seat on the Supreme Court, argued on be-half of social justice as against individual rights based upon property and for the establishment of narrow personal policy. With his "clear and present danger" doctrine, Holmes said that basic liberties were not absolute; they depended upon specific circumstances. He insisted that the Constitution, too, was a liv-ing and changing "experiment" subject to changes in society. More and more privacy cases reached appellate courts through-out the country, and in 1948 the federal district court of Wash-ington, D.C., ruled that the right of privacy extended beyond the mere unauthorized use of a person's name or picture; it extended to *any* unreasonable intrusion into an individual's life.[45]

The legalization of privacy moved like a brush fire through this century, but not without intensifying the debate over what kinds of individual protection are to be accorded and by what means. The debate is plagued with many different premises, objectives, and standpoints. As is apparent in the next chapter, the ongoing effort to secure privacy in the law has been frus-trated, first through the struggle to define what it is we mean by the term, and second by the trial and error method of applying existing laws or creating still more. Samuel Warren and Louis Brandeis started the fire, but their impact has not so much been in the power of their argument as in the high status it gave to privacy and to privacy as a legal right.

In this chapter we have seen how a natural phenomenon became a legal right, how a personal matter acquired the official status of law. As catalysts, Warren and Brandeis, annoyed by an unseemly press, recognized in privacy a moral and spiritual dimension and emphasized in the literature such terms as "inviolate personality" and "individual dignity." In the years following their treatise, legal privacy become part of evolving liberalism and the corresponding prominence given to the individual in society. While the era that gave birth to the right of privacy has faded, the interest among all segments of society has heightened considerably, as the next chapter explains.

3

False Promises, Myriad Objectives

"Privacy, like an elephant, is more readily recognized than described," writes J. B. Young from the British perspective.[1] Raymond Wacks, another resident of the United Kingdom, where privacy is accorded no explicit legal status, also employs the jungle as metaphor: "Privacy has grown so large that it now threatens to devour itself."[2] There is much to be said on the subject, and Young and Wacks have summarized the problem better than most commentators. For, as this chapter illustrates, the enormous amount of both popular and scholarly attention given privacy has not always resulted in meaningful solutions. Because the United States is the nation with the greatest preoccupation with privacy, the reader is asked to endure yet another analysis of its many promises and objectives. Scholars from nearly every academic discipline have addressed the topic.

William Prosser, when dean of the law school at the University of California at Berkeley, published in 1960 what is considered the second most important tract on privacy.[3] Dean Prosser elaborated on Warren and Brandeis's private-facts tort and, based upon more than two hundred court cases, found three other distinct categories: intrusion, false light, and appropriation. His was the first effort to reduce privacy law to manageable proportions. But, because Prosser's torts are based on judicial decisions, rather than on some general concept of privacy, they are limited in scope.

Edward J. Bloustein, on the other hand, expanded and extended privacy to an "interest in preserving human dignity and individuality."[4] His concept, although laudable, imposes an incredible burden on a juridical system that normally relies upon precision. Somewhere between Prosser's specific torts and Bloustein's broad concept are Alan F. Westin and the many writers who have adopted Westin's working definition.[5] His privacy is more narrowly defined as " . . . the claim of individuals, groups, or institutions to determine for themselves when, how, and to what extent information about them is communicated to others." He also includes "the voluntary and temporary withdrawal of a person from the general society through physical or psychological means . . . in a condition of anonymity or reserve." Still another concise definition is offered by the U.S. Office of Science and Technology, an arm of the White House.[6] Privacy is described as "the right of the individual to decide for himself how much he will share with others his thoughts, his feelings, and the facts of his personal life."

Dean Prosser incorporated the four torts in his influential *Handbook of the Law of Torts* and *Restatement (Second) of Torts*, the standard guides for lawyers and jurists.[7] Prosser defined the Warren and Brandeis private-facts tort this way:

> One who gives publicity to a matter concerning the private life of another is subject to liability to the other for invasion of his privacy, if the matter publicized is of a kind that a) would be highly offensive to a reasonable person, and b) is not of legitimate concern to the public.

Of the four, this is the most troublesome for journalists, who must decide when publicity is unreasonable or even when facts must be viewed as private. Although cases such as *Sidis* are troublesome, the courts tend to defer to the news media's interpretation of "legitimate concern to the public."

Intrusion, the second tort Prosser discerned, occurs when an expectation of seclusion is breached, normally the invasion of someone's physical solitude. This category, related to common law trespass, is what most people think of when they discuss privacy invasion, although the protection of private information is now also a widespread concern, and is discussed in a later chapter on computer data collection and dissemination. Strong statutory law protects against such intrusive devices as wiretapping, bugging, and photography.

False light, the third tort, involves the publication of erroneous information that creates a wrong impression about an individual's life or behavior. The information may not be unflattering, but it nonetheless injures personal dignity. The offense is often unintentional or unplanned. An aspect of this category, fictionalization, is something a dramatist, for example, might do intentionally to make the story more appealing. Whether intentional or not, the publication or broadcast of false information is considered an invasion of personal privacy.

Appropriation, the oldest and least ambiguous of Prosser's torts, is the unauthorized use of one's name or likeness for commercial purposes. The first application of the Warren-Brandeis concept was a case involving the use of a young woman's picture in an advertisement without her consent. (See Chapter 2.) She lost, but the case led to the passage of the nation's first privacy law. As part of the appropriation tort, the so-called "right to publicity," which is discussed later as the protection of celebrated personalities, has become nearly a fifth recognized privacy tort. It started to gain acceptance in 1953, with Judge Jerome Frank's assertion that certain individuals, particularly celebrities, had a property right in their name just as they have ownership in land or copyright in a book.

Bloustein's contribution to the promises and objectives of privacy law was the direct result of Prosser's reductionism, which Bloustein argued was too formalistic and missed the essential point of the Warren-Brandeis treatise, plus many of the subsequent right to privacy rulings. His view, based, like Prosser's, upon an extensive review of legal studies and actual cases, is that "our law of privacy attempts to preserve individuality," our "liberty as individuals to do as we will," our "human dignity," by placing sanctions on, as Bloustein put it, "outrageous or unreasonable violations of the conditions of its sustenance." Human dignity, not commercial interests, is for Bloustein the dominant theme underlying the right. Few courts have followed his analysis, for if these desires were fully adopted there might arise a loss of free speech and free press.

In his *Restatement,* Prosser requires the plaintiff to show that the information was in fact private and that the defendant spread the information widely. The action may be defeated by a showing that the public has a legitimate interest in the information. Therefore, the private-facts tort usually involves the mass media because of this mass-publication requirement. Bloustein rejects Prosser's reputational and emotional distress arguments and says that the tort, in preserving some "right to be let alone," really protects individual dignity and integrity and prevents the loss of individual freedom and independence. Bloustein supports the mass-publicity requirement, however. He believes that individual dignity is damaged not when friends, who are assumed to know better, learn things that may change their opinion of us, but when we are "made a public spectacle." The difference is that private gossip has "a kind of human touch and softness," according to Bloustein, and its effect is moderated by the tendency of some listeners to "know and love or sympathize with the person talked about." Newsprint and telecasts, in contrast, are "cold and impersonal." But, as Diane Zimmerman observes, such conclusions are questionable because exposure of private information to people who have no special interest or first-hand knowledge is likely to be more damaging than information circulated widely among strangers by the media. How we are viewed by people we know is likely to be of greater concern than is the opinion of strangers.[8] In this regard, it is noteworthy that in the Sidis case, which is at the center of Bloustein's position, the court found that, although the *New Yorker* article was a "merciless" dissection of his life, the portrait that resulted was sympathetically drawn and "instructive."

Westin, author of the largest study to date, *Privacy and Free-*

dom, begins his defense with the premise that, despite obvious differences among cultural norms, it is still possible to describe the general functions that privacy performs for individuals and groups in Western democratic nations. The "functions" are part of what Westin identifies as four basic "states" of individual privacy: solitude, intimacy, anonymity, and reserve. In the state of solitude, the individual is separated from the group and freed from the observation of other persons. Like Thoreau at Walden Pond, he may be subjected to physical stimuli, such as noise, odors, and vibrations, and his peace of mind may continue to be disturbed by the sensations of heat, cold, itching, and pain. Solitude is the place where the individual communes with himself or with some supernatural force, conducting in Westin's phrase, "that familiar dialogue with the mind or conscience." Solitude is the ultimate state of personal privacy.[9]

In Westin's intimate state, the individual acts as part of a small unit that claims and is permitted to exercise "corporate seclusion" in order to achieve "a close, relaxed, and frank relationship between two or more individuals." Typically, such units include husband and wife, the family, a circle of friends, or certain work-place relationships. "Whether close contact brings relaxed relations or abrasive hostility depends on the personal interaction of the members," writes Westin. "But without intimacy a basic need of human contact would not be met."[10] Whereas solitude is a natural and solitary experience, the state of intimacy is at risk without the understood trust of others who also have a stake in the relationship. Common law already protects the confidentiality of many such arrangements, for example, those between priest and confessant, physician and patient, attorney and client, and husband and wife. Legislated law recognizes similar intimacies, such as that between a journalist and an undisclosed source of news. But the promise of the intimate state is not so much a matter of the right to privacy as it is a matter of understood trust and confidence.

In a state of anonymity, Westin's third category, the individual may appear in public places or perform public acts but

still seek, and find, freedom from identification and surveillance. As Westin describes the anonymous person:

> He may be riding a subway, attending a ball game, or walking the
> streets; he is among people and knows that he is being observed;
> but unless he is a well-known celebrity, he does not expect to be
> personally identified and held to the full rules of behavior and role
> that would operate if he were known to those observing him.

The individual merges into the "situational landscape," and the
sense of relaxation and freedom that one may seek in open
spaces and public places is destroyed by the knowledge or fear
that one is under systematic observation or surveillance.

Anonymous relations also give rise to what Georg Simmel
labeled the "phenomenon of the stranger," the person who
"often received the most surprising openness—confidences
which sometimes have the character of a confessional and which
would be carefully withheld from a more closely related person."[11] In an anonymous environment, the individual can express himself clearly and honestly, as with a bartender, because
he knows that the "stranger" is passing through his life. And,
although the stranger's responses may be objective, he exerts no
real authority over the person in this state of "public privacy."
Again, as with Westin's other states of privacy, the promise of
anonymity is more descriptive than it is prescriptive of a legal
right. Custom, rather than law, is what makes anonymity work
in public surroundings. Although the desire for public privacy is
indeed natural and human, such contradictions are beyond the
force of legal control in an open society.

Reserve, which Westin defines as the most subtle state of
privacy, is the creation of a psychological barrier against unwanted intrusion, as when we feel the need to limit information
about ourselves with the help and discretion of people around
us. "Even in the most intimate relations," Westin writes, "communication of self to others is always incomplete and is based on
the need to hold back some parts of one's self as either too
personal and sacred or too shameful and profane to express."
Simmel called this circumstance "reciprocal reserve and indif-

ference," the relation that creates "mental distance" to protect the personality. It is similar to "social distance" and operates under rules of social etiquette. Simmel identified the tension that often results within the individual as being between "self-revelation and self-restraint" and, in society, between "trespass and discretion."[12]

Westin's states of privacy are connected to four functions that privacy performs for individuals in democratic societies: personal autonomy, emotional release, self-evaluation, and limited and protected communication. They are strikingly similar to Professor Thomas I. Emerson's "values" essential to the maintenance of a system of free expression, the concept with which personal privacy is most often in conflict. Emerson's democratic values are: individual self-fulfillment, the pursuit of truth, participation in social and political decision-making, and the maintenance of both change and stability in society.[13] Westin's functions and Emerson's values are similar means toward the same end of liberal democracy, but the "functions" of privacy and the "values" of free expression are so frequently in conflict that any resolution usually rests in the balancing of priorities. In this debate over personal privacy in an open society, we might heed the advice of Justice Benjamin N. Cardozo, who called freedom of thought and speech "the matrix, the indispensable condition, of nearly every other form of freedom."[14] Privacy, when viewed in this larger context, may be at best a right derived from the collective matrix, where individual rights are subordinate to public interest.

Personal autonomy has its roots in the fundamental belief in the uniqueness of the individual, according to Westin, "in his basic dignity and worth as a creature of God and a human being, and in the need to maintain social processes that safeguard his sacred individuality."[15] Psychologists and sociologists link this sense of the individual to the human need for autonomy and periodic anonymity—"the desire to avoid being manipulated or dominated wholly by others." Social theorists at the center of traditional liberalism seek to defend individual claims against general societal interests. They say the individual is inviolable, and they believe the "core self" is protected by "zones" or "regions" of privacy. And autonomy is threatened by those who penetrate the inner zone.

Emotional release, the second Westin function performed by privacy, stems from the need the individual has to seek relief from physical and emotional stress. Social theorists remind us that life demands a variety of roles, and no person can successfully play these different parts without some relief. In this also, privacy shields the individual from having always to comply with social norms. As Westin sees it: "If there were no privacy to permit society to ignore these deviations—if all transgressions were known—most persons in society would be under organizational discipline or in jail, or could be manipulated by threats of such action."[16] Privacy, in short, allows us to be ourselves at low risk.

Self-evaluation is also enhanced through privacy. In the state of solitude, a person is able to evaluate information that bombards him or her daily in order to act as appropriately and as consistently as possible. Privacy provides the time to anticipate problems, recast doubts, and originate solutions, according to Westin. Theologians, philosophers, and poets have always affirmed man's need for contemplation. Persons must learn to live with themselves before they can live with others.

Limited and protected communication, Westin's final "function," provides the individual with the opportunities he needs for sharing confidences and intimacies with those he trusts. "A friend," said Ralph Waldo Emerson, "is someone before . . . [whom] I may think aloud."[17] Common law recognizes the need to protect certain confidential relationships that can flourish in the state of intimacy. Psychological distance, as respected in a successful marriage, is as important as spatial distance.

Individualism *qua* privacy has long been a cherished American precept; it came with the territory. On the importance of privacy for political liberty, the historian Clinton Rossiter noted:

> The free man is the private man, the man who keeps some of his thoughts and judgments entirely to himself, who feels no over-riding complusion to share everything of value with others, not even those he loves and trusts.[18]

Alan Westin's enormously helpful work is part of the liberal rights-based tradition, which treats personal privacy as the basic instrument for achieving individual goals. Yet, quite obviously, there is another view with an opposite perspective and different goals. It was suggested by John Donne more than three centuries ago, at the beginning of the modern era: "No man is an island, entire of itself; every man is a piece of the continent, a part of the main." In one of his masterful poems, "An Anatomie of the World," Donne envisioned the earth nearing its end:

> 'Tis all in peeces, all cohaerence gone;
> All just supply, and all Relation:
> Prince, Subject, Father, Sonne, are things forgot,
> For every man alone thinkes he hath got
> To be a Phoenix, and that then can bee
> None of that kinde, of which he is, but hee.[19]

Later, Hegel was to advise harmony and coordination in the interest of civil unity, but allowing for the aspirations of individuals to be fulfilled through what he called a "chain of social connexions."[20] In a later chapter we suggest that the Westin states and functions of privacy could be accommodated in a society built upon a sense of community. But here, before exploring other promises and objectives, it is worth noting that the philosophy of individualism found its way into American constitutional law in the twentieth century, as the Supreme Court, in particular, began to elevate the position of personal privacy in an increasingly impersonal world.

One of the more imaginative solutions to the problem of defining and protecting privacy is Gary L. Bostwick's "taxonomy," the basic premise for which is that privacy has become a catch-all phrase, "protecting too little because it protects too much." Where Prosser described privacy by the ways it is illegally invaded, Bloustein by its one basic inviolate element, Westin by its many descriptive elements, Bostwick insists privacy encompasses three separate, and distinct, "rights." Where the others identified invasions, states, functions, and conditions of privacy, Bostwick believes that his trio of rights should be viewed more discretely and analyzed within their appropriate framework. The privacy rights are: repose, sanctuary, and intimate decision. Bostwick's approach, similar to Dean Prosser's four torts, is reductionist in that it tries to contract the expanding concept.[21] It is also similar to the even more generic identifications advanced by P. Allan Dionisopoulos and Craig R. Ducat: place-oriented privacy, person-oriented privacy, and privacy as it inheres in certain human relationships.[22] Bostwick's privacies of repose and sanctuary are place-oriented concepts, and his privacy of intimate decision is person-oriented and relational.

Bostwick asserts that, because even a partial list of recent court cases in which privacy was invoked includes such a variety of circumstances, often disparate in their content, the privacy rubric has become too broad. If privacy is not better described, he maintains, "a right without description is a right without protection."[23] Such a list might include whether a city may refuse to allow political advertising on its buses, whether a father has the right to order that life support systems be disconnected from his comatose daughter, whether a city can restrict the number of unrelated individuals living in one house, and whether evidence is admissible if it results from a search founded upon a warrant granted because of the positive reaction of marijuana-sniffing dogs to a trailer parked in a public space.[24]

All were deemed privacy cases. Bostwick cites them to show that there has been little agreement over the source of privacy rights; all usually were related to Fourth Amendment protection against unreasonable search and seizure. Little else binds them, however. By dividing privacy into "rights," perhaps greater legal protection may be accorded each, and all evaluated more systematically, more objectively. Bostwick's contribution to the growing discussion begs for serious attention here and in a later examination of privacy as a natural or bestowed right.

The privacy of repose, according to Bostwick, is the freedom from anything that disturbs or excites, the opposite of calm, peace, or tranquility. Sometimes the right of repose is in conflict with other protected rights—"freedom from being talked to can be guaranteed only if others are prevented from talking." In *Martin v. Struthers* (1943), the First Amendment prevailed over the right to privacy, "unless the privacy claim was staked out by means of a sign."[25] Associate Justice Hugo Black noted that, although door-to-door solicitation can be a nuisance and perhaps even regulated, it cannot be completely banned in deference to constitutional free speech. In a similar vein, the Supreme Court said that "the living right to privacy and repose" was not to be trampled by opportunists seeking private gain.[26] Privacy, as repose, may only enjoy a modicum of status outside the home, however.[27] Where privacy is place-oriented, the critical element for protection appears to be the degree of captivity. The high court reasoned in another case that the screen at a drive-in theater was not so obtrusive that people could not avert their eyes from nudity.

Bostwick's privacy right associated with sanctuary is meant to prohibit other persons from seeing, hearing, and knowing. Where the zone of repose serves to block out unwanted stimuli, the sanctuary zone attempts to keep things within its boundaries. Both zones have their roots in traditional property rights. The Fourth Amendment is the customary protection of sanctuary, although in the famous *Olmstead* decision in 1928, the Supreme Court ruled that police wiretapping was not unreasonable because nothing tangible, or physical, had been taken or intruded upon.

In subsequent decisions, the court tried to address the question of what expectations of privacy are constitutionally "justifiable," that unreasonable searches and seizures are protected, that people have a right to expect a reasonable amount of privacy.[28] But, where public interest is compelling, the courts may be expected to prefer freedom of speech and of the press. Also, if private information is obtained from public records, which are open to the public, privacy is expended.

Bostwick's third right, the privacy of intimate decisions, is more dynamic than either of the others and implies less "freedom from" and more "freedom to." Here the landmark is *Griswold v. Connecticut* (1965), the court's invalidation of the state's birth-control law as a violation of marital privacy.[29] In this decision, the justices made a strong constitutional statement on at least the yearning for privacy. "Various guarantees create zones of privacy," Associate Justice William O. Douglas wrote for the majority. But, whereas personal privacy was endorsed in *Griswold*, the court also based its decision on the public's right to know "available knowledge," setting up a potential conflict between private rights and public interest in future privacy cases. Douglas included the right of association as part of the First Amendment and noted that other facets of privacy are contained in the Third, Fourth, and Fifth Amendments, as well as in the Ninth, for good measure. Other examples of intimate-decision privacy, that is, privacy adhering in the person, are the controversial abortion decisions of 1973, which, because the rulings focused on privacy, bear upon a later discussion.

Bostwick, meanwhile, summarizes his privacy rights as follows: "Repose maintains the actor's peace; sanctuary allows an individual to keep some things private; and intimate decision grants the freedom to act in an autonomous fashion."[30] The real value of this approach, like Dean Prosser's torts, is that it is based upon case studies, the many actual situations with which the courts have grappled and either allowed or disallowed constitutional protection. Thus, Bostwick's study is objective, but his many illustrative cases do suggest that the judiciary has been forced into a more philosophic posture than ever before.

In *Katz v. U.S.* (1967), the Supreme Court introduced the element of legitimate "expectation of privacy" and attempted to define further what it has since called "constitutionally protected areas" for private behavior. Associate Justice John M.

Harlan wrote that the Fourth Amendment protects people, not places, explaining that "there is a twofold requirement, first, that a person have exhibited an actual (subjective) expectation of privacy and, second, that the expectation be one that society is prepared to recognize as 'reasonable'." The Court held in *Katz* that electronic listening in on a telephone conversation in a public telephone booth violated privacy as protected by the Fourth Amendment's search and seizure provision. Justice Harlan likened the telephone booth, although a public facility, to a house, where there is more expectation of privacy than in, for example, a motor vehicle. He said there is less expectation of privacy in a motor vehicle than a house, however. The Court warned, "What a person knowingly exposes to the public . . . is not a subject of Fourth Amendment protection." Justices Black and Potter Stewart, in their provocative dissents in *Griswold*, presented later, also questioned the extent to which the privacy right can be covered by the Constitution.

As with the approaches to privacy protection advanced by Prosser, Bloustein, and Westin, Bostwick's falls victim to a descriptive profile of personal privacy. His study provides little of prescriptive value, unless it is in the observation that society's fluid quality allows for changes in our ideas about privacy and, hence, our inability to stabilize the contours of zones of privacy. Like J. B. Young's elephant, privacy is more readily recognized than described, but Bostwick's approach comes closer than most.

Several nonjuridical academic disciplines have also developed a research interest in privacy. Prominent among them is the field of communication studies, which has produced a number of insights on what Judee K. Burgoon calls "dimensions" of privacy.[31] Communication scholars ask such penetrating questions as: What distinguishes private from nonprivate? How essential is privacy to individuals and to groups? How is privacy expressed or manifested? How does one achieve or restore privacy? Is privacy a major goal of our society?

Of more than passing importance, too, is the fact that communication researchers are far less concerned with legalistic notions of privacy than they are with, as Burgoon aptly writes, the "profound implications of privacy loss for interpersonal transactions." Relying upon an increasing variety and number of controlled studies, scholars have related communication issues to privacy concerns—reticence, verbal and nonverbal behavior, shyness, loneliness, and unwillingness to communicate, to name but a few of the human elements involved. Privacy, in these terms, is the individual seeking and then maintaining a relative degree of freedom from contact, surveillance, or social pressure. The results of communication studies may not bring us closer to a better understanding of legal, or even moral, privacy, but they force us at least to reconsider some basic elements of the complex human condition.

Burgoon, who recently reviewed current research findings, identifies four broad dimensions: physical privacy, social privacy, psychological privacy, and informational privacy. Physical privacy is the degree to which one is physically inaccessible to others. She says that physical privacy is necessary for the optimal functioning of the individual within society and for society itself. Privacy, thus, is individual and subjective. Social privacy is often the consequence of the physical dimension and is similar to Westin's notion of limited and protected communication. This kind of privacy allows for "situational exigencies" and individual differences. Overlooking such jargon that sometimes creeps into these assessments, the idea is that the social dimension is supposed to permit unusual personal behavior without undue fear of punishment by the larger community. Withdrawal into privacy is the "palliative of unbearable relationships."[32] As expressed by one communication scholar: "It is a necessary condition of privacy (though not sufficient to define 'privacy') that the individual is free from the power or influence of other people."[33]

Moving through the physical to the social and then to the psychological dimension of privacy, it is clear that not all translate easily into "rights" for juridical decision-making. Physical privacy, the most obvious, has long been protected at common law and, in the case of electronic and photographic eavesdropping, by constitutional and legislated law as well. Social privacy is somewhat less protected, as is psychological privacy, for, as Burgoon affirms, the degree of privacy is dependent upon each person's interpretation of any given situation. These dimensions are an extra-legal aspect of the private life. Psychological privacy, by its nature, pertains only to the individual and not to groups. Although it is true, in the words of the German sociologist Georg Simmel, that a violation of psychological privacy "effects a lesion of the ego in its very center," it is also a matter best left to individual discretion.[34] If a society does not value by custom self-fulfillment and self-realization, neither is it capable of forcing such norms on individuals by legal means. George Orwell blamed official society and the mass media, especially television, for the end to private life. But Aldous Huxley before him attributed that chilling loss more compellingly to personal choice and collective apathy. In *Brave New World*, Huxley prophesied a society controlled not by Big Brother, but by pleasure.

Informational privacy, the fourth dimension with implications for interpersonal communication, is the most talked about and the most problematic aspect of privacy today. It is what most people have in mind when they speak of privacy intrusion. It is the focus of casual debates and the topic of professional seminars. Most books on the question classify personal information as personal property. Closely allied with the psychological dimension, informational privacy also has a political dimension that transcends ideology. One definition is widely held: "The right of an individual to determine how, when, and to what extent data about oneself are released to another person."[35]

The Council of State Governments amplified the definition at a seminar on privacy rights:

> Protection of privacy is largely concerned with assuring that accurate, relevant, and timely information about people is used only

for stated purposes and in the best interests of each individual. It includes giving the individual both control over how information about him is used and a mechanism for making corrections to the record.[36]

Most such statements see data collection and data use as serious threats to personal privacy by governmental agencies, political organizations, and private industry. Modern technology has made eavesdropping easy.

Each of Burgoon's dimensions are clearly interrelated, although in their own way each also has different implications for human communication. These interpersonal dimensions are significantly different from the several anthropological states and functions of privacy or the various juridical torts, and they help us to understand the broad nature of privacy. Also, research findings imply a variety of ways, or strategies, that individuals and groups have employed in maintaining or restoring privacy. Communication scholars are primarily interested in the actual ways people cope with complex modern society. Legal scholars too frequently rely upon the ways people ought to act or might act regarding, in this instance, their individual or collective perception of privacy. Communication scholars attempt to learn what people do "in their everyday habitats and everyday lives."[37] Their empirical studies have certain limitations, yet the very identification of message strategies and coping mechanisms helps to create a rationale for achieving solitude and seclusion through extra-legal means. Communication studies come closest to describing beyond mere legal recognition J. B. Young's mammoth beast.

Of the four communicational dimensions outlined by Burgoon, only informational privacy is clearly beyond the exclusive control of the individual. Thus, in order to reduce the threat of invasion it has been necessary for federal and state legislatures to enact appropriate measures. As for the other dimensions, however, studies have shown a high degree of individual self-control over physical, social, and psychological privacy. It would seem to follow, therefore, that legal protection is unnecessary, if not irrelevant.

Privacy, whether considered as a moral entity or as a legal right, has suffered from too much rhetorical attention since Warren and Brandeis elaborated on Judge Cooley's simple but vague legal concept, "the right to be let alone." As implied earlier in the chapter, efforts to define "privacy" or "invasions of privacy" have caused the original idea to grow and become unwieldy. Cooley's concept was at once vague and comprehensive. Most of the attention amounts to normative or descriptive statements—what the author would like privacy norms to be or how the courts have interpreted the common law and the Constitution.

But there are differences, too, among the various functions, states, and dimensions of privacy. Prosser and Bostwick describe the application of law to personal autonomy, property, and other liberties that have come to be associated with privacy. Prosser's torts become Bostwick's generic rights in translation. Not satisfied with such simplifications, Bloustein insists that there is a single interest behind the law's protection, what he calls "human dignity." Westin wants a narrow but absolute "claim" to control individual, group, and institutional information. His approach is both normative and descriptive.

Louis Lusky, in a thoughtful critique of Westin's claims, suggests that privacy is more a condition than a right. He says we must begin to break down our unitary concept and seek more useful and precise terminology. We must identify those aspects of privacy that are absolute, and separate them from the contingent ones. Lusky believes we can then more readily deal with the legal aspects, for "right" implies legal right, and clearly not all privacy can be legalized.[38] This observation is similar to that of Raymond Wacks, who regrets the continuing colonization of traditional liberties by privacy. Privacy is now entangled with confidentiality, secrecy, defamation, property, and the storage of information. "It would be unreasonable to expect a notion so complex as 'privacy' not to spill into regions with which it is closely related," Wacks writes, "but this process has resulted in the dilution of 'privacy' itself, diminishing the prospect of its own protection as well as the protection of the related interests."[39]

Of the numerous studies Burgoon cites in identifying the several dimensions of communicational privacy, Edward T. Hall's physical distance zones suggest some extra-legal strategies for dealing with privacy, its loss as well as its gain. The zones are: intimate, personal, social, and public. Intimate distance is zero to eighteen inches, close enough to touch, to reach extremities, to distort the visual system, and to detect body heat and odor. Personal distance, eighteen inches to four feet, is more formal, close enough but not too close, limiting the possibility of touch. Social distance, four to twelve feet, occurs usually in impersonal or casual exchanges and reduces sensory involvement. Public distance, from twelve feet and beyond, is reserved for public occasions.[40]

Although privacy does not always correlate closely with actual physical distance, when combined with these zones, the courts have found we acquire some human behavioral grounds for legal protection of, for example, medical records that often include intimate information. Information on a vasectomy may be deemed of intimate distance, except when used for statistical purposes without identifying the patient. Similarly, information on an abortion may be given personal or social protection. In all of Hall's distances, public interest, as even Warren and Brandeis allowed, will have to remain a factor in deciding the degree to which otherwise private matters become public, and thus more difficult to receive protection. For, if the determination were left to individuals, the law simply would become an unmanageable morass—ultimately unprotective. The effect of the early court rulings, as discussed in Chapter 2, and reinforced by scholarly attention, as explained in this chapter, has been the creation of the right not to suffer outside interference with certain private activities. Most of the legal decisions have been inferred from Dean Prosser's torts—but more and more, the grounds have been "found" in the Constitution itself. We now turn to that phase of privacy in a public society.

4

Disagreement on Zones

Although the term "privacy" appears nowhere in the U.S. Constitution, the Supreme Court discovered protection for private activities in the Bill of Rights as early as 1886, four years before Warren and Brandeis published their law review article. The court held, in *Boyd v. United States*, that federal subpoenas for certain business records violated the due process clauses of the Fourth and Fifth Amendments.[1] The Court said that the doctrines of the amendments "apply to all invasions on the part of the government and its employees of the sanctity of a man's home and the privacies of life." The Court went on to identify the "privacies" it had in mind: "It is not the breaking of his doors, and the rummaging of his drawers, that constitutes the essence of the offense; but it is the invasion of his indefeasible right of personal security, personal liberty, and private property." With this, privacy won a permanent place in American jurisprudence.

In *Union Pacific Railway Co. v. Botsford* (1891), the Supreme Court held that a federal trial court lacked authority to order the plaintiff "to submit to a surgical examination as to the extent of the injury sued for." Justice Horace Gray, speaking for the majority, said, "No right is held more sacred, or is more carefully guarded, by the common law, than the right of every individual to the possession and control of his own person, free from all restraint or interference of others, unless by clear and unquestionable authority of law." He wrote that the right to one's person may be said to be a right of "complete immunity: to be let alone," as Judge Cooley and Warren and Brandeis after him had urged.[2]

The Fourteenth Amendment's due process clause came into play in *Meyer v. Nebraska* (1923), when the Court found unconstitutional a state law prohibiting the teaching of any modern language other than English in any private, denominational, parochial, or public elementary school. The Court said that the statute, as applied to instruction in German in a parochial school, "unreasonably infringes the liberty guaranteed . . . by the Fourteenth Amendment," which states that the government shall not deprive any person "of life, liberty, or property, without due process of law."[3]

Education, thus, became another privacy accorded constitutional protection. But the court went beyond that of the specific Nebraska episode to identify other liberties, freedoms, or rights embedded in the right to be let alone:

> While this Court has not attempted to define with exactness the liberty thus guaranteed . . . without doubt, it denotes not merely freedom from bodily restraint, but also the right of the individual to contract, to engage in any of the common occupations of life, to acquire useful knowledge, to marry, to establish a home and bring up children, to worship God according to the dictates of his own conscience, and, generally, to enjoy those privileges long recognized at common law as essential to the orderly pursuit of happiness by free men.

The pursuit of happiness was thus added to the growing list of life's protected privacies.

The Meyer decision affected the outcome in *Pierce v. Society of Sisters* (1925),[4] in which the Court struck down an Oregon statute that required children to attend public rather than private or parochial elementary schools. But the state's argument was not without principle, however, as its counsel asserted:

The voters of Oregon might have felt that the mingling together, during a portion of their education, of the children of all races and sects, might be the best safeguard against future internal dissention and consequent weakening of the community against foreign dangers.

In upholding a parent's right to educational privacy, the Court again preferred individual freedom over a community goal.

Associate Justice McReynolds's majority opinion, similar to what he wrote in *Meyer*, held that the Oregon law "unreasonably interferes with the liberty of parents and guardians to direct the upbringing and education of children under their control." The American view on individualism and individual responsibility is nowhere more forcefully stated than in the justice's phrase:

The fundamental theory of liberty upon which all governments of this Union repose excludes any general power of the state to standardize its children by forcing them to accept instruction from public teachers only. The child is not the mere creature of the state; those who nurture him and direct his destiny have the right, coupled with the high duty, to recognize and prepare him for additional obligations.

And one way, the Court concluded, to help ensure the realization of "liberty" (for the parents) and "destiny" (for the child) is through a right to privacy.

Marriage and procreation became private activities that enjoy constitutional protection in *Skinner v. Oklahoma* (1942).[5] The Court ruled against a state statute that provided for the sterilization of persons convicted three or more times of felonies involving moral turpitude, but excepted those convicted of "offenses arising out of the violation of the prohibitory laws, revenue acts, embezzlement, or political offenses." For the Court, Justice William O. Douglas noted: "We are dealing here with legislation which involves one of the basic civil rights of man. Marriage and procreation are fundamental to the very existence and survival of the race."

The pursuit of "lawful private interest privately" was supported by the court in *NAACP v. Alabama* (1958), when it held unconstitutional a state order that required the association to produce membership lists.[6] Justice John M. Harlan, writing for the Court, said:

> The immunity from state scrutiny of membership lists which the Association claims on behalf of its members is here so related to the right of the members to pursue their lawful private interest privately and to associate freely with others in so doing as to come within protection of the Fourteenth Amendment.

Associational privacy thus became a protected right.

Two dissenting opinions by Supreme Court justices have had considerable influence on the development of the protected privacies of life. First, Justice Brandeis objected to the majority's refusal to treat wiretapping as an invasion of Fourth Amendment rights in *Olmstead v. United States* (1928).[7] His view has been widely quoted.

> The makers of our Constitution undertook to secure conditions favorable to the pursuit of happiness. They recognized the significance of man's spiritual nature, of his feelings and of his intellect. They sought to protect Americans in their beliefs, their thoughts, their emotions and their sensations. They conferred, as against the Government, the right to be let alone—the most comprehensive of rights and the right most valued by civilized men.

Brandeis had picked a sympathetic forum in which to restate the concept he and Samuel Warren had published thirty-eight years earlier. But it is nonetheless presumptuous, if not historically inaccurate, to revise Judge Cooley's original right—"immunity from attacks and injuries"—and claim it as the "most comprehensive" of human rights. The Court eventually overruled *Olmstead*, but it did so on more specific, and substantive, provisions of the Constitution.

The second important dissent, which has been a major influence in subsequent privacy decisions, is Justice Harlan's in *Poe v. Ullman* (1961).[8] The case dealt with two married women and a doctor who had sought judgments declaring unconstitutional

Connecticut statutes prohibiting physicians from giving contraceptive advice. In dismissing their appeal, the Supreme Court noted that the state's history of failure to enforce its ban did not indicate a serious threat of prosecution. Harlan objected to the Court's refusal to deal with the broader constitutional issues and said that the law violated the Fourteenth Amendment.

> I believe that a statute making it a criminal offense for *married couples* to use contraceptives is an intolerable and unjustifiable invasion of privacy in the conduct of the most intimate concerns of an individual's private life.

As if to anticipate the Court's turnabout in *Griswold v. Connecticut* four years later, Harlan said that the reach of the Fourteenth Amendment due process clause was not limited to the first eight amendments, but, rather, privacy constituted a fundamental right. "It is a rational continuum which, broadly speaking, includes a freedom from all substantial arbitrary impositions and purposeless restraints." Again, a specific instance of invasion was viewed as the reason for broad protection. Harlan's dissent created the biggest umbrella yet for privacy, which the majority adopted in *Griswold* but whose due process approach in that case even appeared too liberal until it was accepted in the controversial abortion decisions of 1973.

Meanwhile, when the Court overruled *Olmstead* in 1967, Congress enacted legislation the next year regulating all forms of wire and electronic surveillance by law enforcement agencies.[9] The Omnibus Crime Control and Safe Streets Act of 1968 prohibited such activity without court orders. In a subsequent case, *U.S. v. U.S. District Court* (1972), the Court noted that official surveillance risks infringement of constitutionally protected "privacy of speech."[10] Other activities that the Supreme Court has accommodated under the aegis of "privacy" include long hair and advertising.

With *Griswold*, the Supreme Court made its clearest statement to date, as well as its most sweeping, of the constitutional foundations of, in the words of one law professor, "a yearning for privacy." The decision invalidated a Connecticut law, upheld in

Poe, forbidding the dissemination of birth control information as a violation of a right to marital privacy. Justice Douglas, delivering the opinion of the seven-member majority, said that any important liberty not safeguarded by the Bill of Rights can be found in the "penumbra," or shadow, of a specific guarantee. "Various guarantees create zones of privacy," he wrote, and cited the right of association, for example, as part of the First Amendment. He noted other facets of privacy contained in the Third, Fourth, and Fifth Amendments, and included the Ninth for good measure: "The enumeration in the Constitution, of certain rights, shall not be construed to deny or disparage others retained by the people."

Justice Harlan concurred but did not join the majority completely. The penumbra approach, although sufficient to strike down the Connecticut statute, did not go far enough, and he found that the Court's "incorporation" doctrine could be used later to restrict the Fourteenth Amendment due process clause.

> For me this is just as unacceptable constitutional doctrine as is the use of the "incorporation" approach to *impose* upon the States all the requirements of the Bill of Rights as found in the provisions of the first eight amendments and in the decisions of this Court interpreting them.

Harlan, in differing dramatically with Douglas, said that the decision need not depend on "radiations" from the Bill of Rights: "The Due Process Clause of the Fourteenth Amendment stands . . . on its own bottom." Harlan's position is dramatic because, whereas it recognized the personal nature of marriage, it implied that confusion would probably follow when the courts, like the shepherd boy who cried wolf once too often, are confronted with genuine infringement.

Justices Black and Stewart were the lone dissenters in *Griswold;* Black because he could find no *specific* language in the Constitution protecting a "broad, abstract and ambiguous" right of privacy, and Stewart because he could find no *general* rights of privacy in the Bill of Rights or in any part of the Constitution. Their views, which are especially insightful, deserve attention because they are supportive of the thesis that the right to privacy has grown too large to be meaningful, or even effective. It threatens to devour itself, but also, and more importantly, as a constitutional right it obviates common law and circumvents public legislated law. It defies tradition.

Black admitted at the outset that his view had nothing to do with the belief that the Connecticut law was wise or that its policy was a good one. He said that had the defendants done nothing more than express opinions to persons coming to the clinic for birth control information, then clearly they would be protected by the First and Fourteenth Amendments, which guarantee freedom of speech. "But speech is one thing; conduct and physical activities are quite another." As active participants, the defendants performed physical examinations and engaged in conduct as well as speech.

Black tried to raise the level of discourse by addressing the philosophical issue:

> One of the most effective ways of diluting or expanding a constitutionally guaranteed right is to substitute for the crucial word or words of a constitutional guarantee another word or words, more or less flexible and more or less restricted in meaning.

This, he said, is the case with the term "right of privacy" as a comprehensive substitute for the Fourth Amendment's protection against "unreasonable searches and seizures." Privacy is a "broad, abstract and ambiguous concept which can easily be shrunken in meaning but which can also . . . easily be interpreted as a constitutional ban against many things other than searches and seizures."

Well-known for his absolutist views on most First Amendment decisions, Black wrote that freedoms protected by that amendment have suffered from a failure of the courts to stick to the simple language expressed, instead of "invoking multitudes of words substituted for those the Framers used." In an important footnote to his dissent, Black alluded to the Warren and Brandeis article and opined that the Court now appeared to exalt the "right of privacy" phrase to the position of a constitutional rule that would prevent legislatures from enacting any law deemed by the Supreme Court to interfere with privacy. "I like my privacy as well as the next one," Black wrote, "but I am nevertheless compelled to admit that government has a right to invade it unless prohibited by some specific constitutional provision."

Black found that the Griswold decision lacked sufficient specificity. He invoked the view of the late Judge Learned Hand, who believed that judges should not use the due process formula or anything like it to invalidate legislation offensive to their "personal preferences." Judge Hand once made the statement, with which Black fully agreed: "For myself it would be most irksome to be ruled by a bevy of Platonic Guardians, even if I knew how to choose them, which I assuredly do not." Harlan, on the other hand, insisted, in dissent in *Poe* and with the majority in *Griswold*, that the courts had the power to strike down laws they consider arbitrary and unreasonable. In this, the justice affirmed the court's long-standing doctrine of judicial review. Since as early as 1803, the high court has had the right of review over the democratically elected branches of government, but its use has nearly always been influenced by political currents and social necessity. With *Griswold*, privacy assumed the clear status of a constitutional right, the product of Supreme Court power as coaxed by the rising concern for human rights in the 1960s. Harlan's position prevailed, as it did in the later abortion decisions.

Justice Stewart, in his dissent, called the Connecticut statute "an uncommonly silly law," which was "obviously unenforceable." As a philosophical matter, the use of birth control devices in marriage should be left to personal and private choice, he said, "based upon each individual's moral, ethical, and religious beliefs." As a matter of social policy, Stewart said that professional advice on birth control methods should be available to everyone. But, in *Griswold*, the Court was not asked to decide whether the law was unwise or asinine. "We are asked to hold that it violates the United States Constitution. And that I cannot do."

In dismissing Douglas's penumbra argument, Stewart observed correctly that there was no "general right of privacy" in the Bill of Rights, nor in any other part of the Constitution, nor, he asserted, "in any case ever before decided by this Court." This, too, was accurate, for in earlier decisions private activities were accorded the protection of specific amendments. Basing his solution to the problem upon the statute's failure to conform to "current community standards," as noted in oral argument before the Court, the justice advised that "the people of Connecticut can freely exercise their true Ninth and Tenth Amendment rights to persuade their elected representatives to repeal it. That is the constitutional way to take this law off the books." This was precisely the subsequent action taken by Connecticut legislators.

Two days after the court handed down its decision, the *New York Times* criticized Justices Black and Stewart.

> A reasonable and convincing argument can be made—and was made by the dissenters—that this infringement on personal freedom represented in the laws of Connecticut and many other states should have been corrected by the legislatures. But the fact is that it was not corrected. To what forum but the Supreme Court could the people then repair, after years of frustration, for relief from bigotry and enslavement?[11]

Yet, from a careful reading of the *Griswold* opinion, as well as from analyses, it is difficult to conclude, as did the newspaper, that bigotry and enslavement were juridical matters before the Court. According to one recent observation: "It was a put-up job, a test case in which the players knew exactly what roles they were to play and were willing to take the risk to fight for what they believed in." It was an overt act of civil disobedience for Estelle T. Griswold, executive director of the state's Planned Parenthood League. "When . . . she opened the doors of a birth control clinic . . . she was begging to be arrested."[12]

The defendants, whose desires had been rebuffed four years earlier, had decided the time was ripe to challenge the law again. Justice Felix Frankfurter, who wrote in *Poe* that the high court "cannot be umpire to debates concerning harmless, empty shadows," had retired in ill health and had been replaced by Byron White. Because the previous court had not been persuaded by the First Amendment argument, the *Griswold* briefs assigned a higher role to the privacy question. Professor Thomas I. Emerson of the Yale Law School, who argued the case, said in an interview: "What is involved really is a right to keep the government out of a certain zone of activity, particularly dealing with sexual matters and intimacy of the home."[13]

A great deal of legal strategy went into the case. One of the attorneys representing Ms. Griswold and the clinic's medical director waived her clients' right to a jury trial because she feared a sympathetic jury. Attorney Catherine Roraback wanted to make sure that there was a conviction on which to base an appeal. When the time came for argument before the Supreme Court, Roraback believed that the right of privacy was relevant, despite denial at the state level that anyone's privacy had been violated because the women who testified had done so voluntarily. Although there is no *specific* mention of privacy in the Constitution, Griswold and company argued that there *ought* to be. Justice Douglas and six of his colleagues took the hint. His angry dissent and that of Harlan in *Poe* had this time resulted in a majority, and the Connecticut law became unconstitutional. The Court had set out on a journey that would broaden even further the ground for a general constitutional protection for personal privacy.

The matter of personal privacy in an otherwise public society— the central paradox running throughout this narrative—is obviously more than just a personal matter. Any matter of concern to the individual is also of equal concern to society in general. Although the focus of this chapter consists mainly of the several important Court rulings that have, in effect, constitutionalized the right to privacy, we should not lose sight of the fact that at any given time there are countless other forces at work influencing society's direction and values. Furthermore, the courts are subjected to the same emerging direction and values. On this, the historian John Lukacs provides an interesting, and not unconnected, series of dates, events, and statistics, which, taken together, set the scene for the Supreme Court's controversial abortion decisions of 1973. Lukacs writes:

> In 1913, in the same year that Mother's Day became a nationally observable holiday, the American people passed another milestone: for the first time in American history more than one person in one thousand was divorced. In that year Margaret Sanger coined the term "birth control." This chronological coincidence of Mother's Day with divorce and birth control is at least telling. By 1930 the American divorce rate rose by another 56 percent, by 1970 more than 350 percent. By 1980 nearly one out of two new marriages was bound to terminate by divorce. The American fertility rate was extraordinary: 3.5 children in the 1950s; this had fallen by one half two decades later. In 1978 alone there were more than a million abortions; in the same year nearly half of American women under the age of twenty-four were childless. The number of unmarried women increased by one third from 1970 to 1980; the number of unmarried couples living together doubled, while the number of children decreased drastically. In 1980, for the first time in the history of the United States, nearly two thirds of American homes (63 percent) did not contain a child.[14]

Lukacs's paragraph is inserted as a reminder of the context, the atmosphere, in which law-making takes place. One could infer, for example, that a substantial change in the *nature* of personal privacy has taken place, from the *figurative* solitude of the eighteenth century to *literal* self-interest of the twentieth century. At least that is one inference. To put it more precisely, as Michael Kammen, another historian, observed in 1971, "The American is more preoccupied with private or personal values than with social or political ones."[15] But this is not to say that marriage, divorce, abortion, rearing children, and the litany of other privacy rights identified and endorsed by the courts are just the result of American self-indulgence. What is suggested is that such private decisions are now condoned, even encouraged, more than ever before.

Prior to 1820, the government was not especially concerned about the practice of abortion. Courts usually followed the English common law, which did not condemn abortion before "quickening," when the mother could feel movement of the fetus. If performed after quickening, abortion was considered a crime. Yet most early state statutes were not aimed at making abortion a criminal act; rather, they were enacted to prevent unsafe practices, poisonous remedies, and incompetent practitioners. As the New Jersey Supreme Court emphasized in 1858, the laws were adopted "to guard the health and life of the mother against the consequences of such attempts."[16] The laws represented a desire by the medical profession to regulate, improve standards, and eliminate unqualified practitioners. After 1880, abortion was illegal in all states, permitted only when the mother's life was endangered by her pregnancy.

Most of the literature published during the second half of the nineteenth century condemned abortion, but the authors conceded that public opinion was not on their side. In the words of one observer in 1891, "Society is indifferent, the Church asleep and the public conscience is dead."[17] Thus, legislation against abortion had largely been the response to pressure from the medical profession, not from popular demand based upon moral or religious belief. Abortion, which carried overtones of shame and guilt, was an old-fashioned private matter, more a matter of personal morality than of personal privacy. Although the decision to have an abortion then, or now, is indeed a private choice, the abortion issue is hardly private—it is very public.

The *Roe* and *Doe* decisions of 1973 legalized abortion, but they also changed the Court's image by fostering renewed attacks on judicial activism and mobilizing both supporters and opponents of abortion.[18] The rulings legalized abortion, but they did not legitimize the policies thus promulgated. Abortion is one of those occasional issues that reflects competing concepts of values and morality, the kind of issue that goes beyond traditional debate over competing economic interests. Abortion has been politicized, but the battle lines are not nearly as clear as they are on most political issues. Normally, but not always, the Supreme Court's use of its acquired right of judicial review has served to alleviate factional disputes. Legislators may not like to have their hard work ruled unconstitutional, but they have learned to live with the vicissitudes of compromise politics. With the abortion issue, as with the broader question of privacy, the Court finds itself quite inappropriately at the center of a worldwide debate on the philosophical issue of personhood.

Brian Barry, a professor of political science and philosophy at the University of Chicago, observed:

> It would be hard to maintain that the history of public discussion of abortion [since the Court's decision in *Roe v. Wade*] has been more rationally conducted in America than in countries where liberalization was carried out through legislative means.[19]

Barry also points out that many people in California, the base of a recent sociological study, were personally opposed to abortion yet accepted the liberalized conditions for obtaining one as long as it was simply a matter of legislative fiat—"a matter of power rather than principle." What outraged, and subsequently mobilized them, Barry believes, was the Supreme Court's decision that abortion was a moral right guaranteed by the Constitution.[20]

On January 22, 1973, the Supreme Court, in a seven-to-two decision, ruled unconstitutional the restrictive abortion laws of Texas and Georgia (and virtually those of every other state). In Texas, it was a crime to abort a fetus or to procure or attempt an abortion except upon "medical advice for the purpose of saving the life of the mother." The Georgia statutes made abortion a crime unless, in the "best clinical judgment" of a licensed physician, abortion was necessary because a continuation of the pregnancy would endanger the life or health of the mother, or the fetus would be born defective, or the pregnancy was the result of rape. A number of procedural conditions were also imposed, including a residency requirement and the approval in advance by a hospital staff abortion committee.

Jane Roe (a pseudonym) challenged the Texas statutes on the ground that they denied equal protection of the law, which placed a disproportionate burden on all but wealthy persons who had the money to travel to places where abortions were legal. She also claimed that she had been denied due process of law because the law was vague and did not spell out just what circumstances came within the ambit of saving the mother's life. Finally, Ms. Roe said the law denied a mother's right of privacy under the First, Fourth, Fifth, Ninth, and Fourteenth Amendments. Mary Doe's Georgia case was argued on similar grounds.

Justice Harry A. Blackmun, writing for the Court, leaned on *Griswold:*

> The Constitution does not explicitly mention any right of privacy. In a line of decisions, however . . . the Court has recognized that a right of personal privacy, or a guarantee of certain areas or zones of privacy, does exist under the Constitution. This right of privacy, whether to be found in the Fourteenth Amendment's concept of personal liberty and restrictions upon state actions, as we feel it is, or, as the District Court determined, in the Ninth Amendment's reservation of rights to the people, is broad enough to encompass a woman's decision whether or not to terminate her pregnancy.

The Court did not agree, however, that this right was absolute, that she is entitled to end her pregnancy at whatever time, in whatever way, and for whatever reason she alone chooses: "We therefore conclude that the right of personal privacy includes the abortion decision, but that this right is not unqualified and must be considered against important state interests in regulation." Among those areas that may be regulated, the Court said, are the state's interest in safeguarding public health, maintaining medical standards, and in protecting potential life.

The decisions did not go as far as some pro-choice advocates would have liked, for the "right" still rested with a doctor having to make a medical decision. In *Roe,* the Court said that a pregnant woman "cannot be isolated in her privacy," making the situation inherently different from marital intimacy, or bedroom possession of obscene material, or marriage itself, or procreation, or education. Although the Court elevated abortion to the status of a privacy right, it did not go as far as an unqualified right. Blackmun posited a possible solution to the bigger issue when he addressed, then sidestepped, the question of when "personhood" begins:

> All this . . . persuades us that the word "person" as used in the Fourteenth Amendment does not include the unborn. . . . We need not resolve the difficult question of when life begins. When those trained in the respective disciplines of medicine, philosophy, and theology are unable to arrive at any consensus, the judiciary, at this point in the development of man's knowledge, is not in a position to speculate as to the answer.

For both sides of the abortion debate, this was the most distressing caveat of all. Personhood was discussed, but the Court relegated it to the realm of opinion, which even pro-choice advocates had been willing to forgo prior to 1973 in the interest of reforming restrictive abortion laws. According to Kristin Luker,

> What was worse, it was defined as an opinion belonging to the *private* sphere, more like a religious preference than a deeply held social belief, such as belief in the right to free speech. It was as if the Supreme Court had suddenly ruled that a belief in free speech was only one legitimate opinion among many, which could not therefore be given special protection by any state or federal agency.[21]

Still, it must also be said that all too frequently the courts have been expected to act in a void created by other segments of society. Perhaps Americans expect too much from their judiciary, despite the occasional use of "raw judicial power" (the view of one dissenting justice), and should look elsewhere for solutions to the nation's moral ambiguity.

Of the numerous treatises, both pro and con, that have been written on the abortion decisions, none is more comprehensive than John T. Noonan, Jr.'s book-length treatment, *A Private Choice* (1979). His frontal attack is at once scathing and thoughtful:

> Each act of abortion is, by declaration of the Supreme Court of the United States, a private decision. Yet each act of abortion bears on

the structure of marriage and the family, the role and duties of parents, the limitations of the paternal part of procreation, and the virtues that characterize a mother. Each act of abortion bears on the orientation and responsibilities of the obstetrician, the nurse, the hospital administrator, and the hospital trustee. The acceptance of abortion affects the professor and student of medicine and the professor and student of law. In the United States, abortion on a large scale requires the participation of the federal and state governments.[22]

In sum, abortion is anything but a matter of privacy. The Court in recent years has been especially protective of the right to privacy, but it has usually done so with the Constitution in mind and, from what may be inferred from the framers' intentions, those unstated but implied protections.

Noonan, a professor of law, believes that the decisions were remarkably harsh, without principle, and an example of naked political preference. He contends that most legal scholars have judged the abortion liberty to lack a constitutional basis. He attributes the rulings to the powerful influence of certain politically potent groups and accuses the seven members of the Court of succumbing, not to a higher juridical principle but to the weight of these groups. And he charges the press with "beating on the drum of bigotry" and for having "effectually misled the American public on the issues at stake in abortion."[23]

Noonan also blames the Court for the political polarization that has come in the wake of the decisions, the justices acting as though their ruling was without social underpinnings and consequences.

The obligation of the government is to aid the disadvantaged by social assistance and economic improvement; the liberty transforms this responsibility to the poor into a responsibility to reduce poverty by reducing the children of the poor.

The "liberty" of abortion, for Noonan, is yet another case study in the law's ability to mask realities of human life with harmful judicial abstractions. "There must be a surpassing of such liberty by love."[24]

John Hart Ely, another constitutional law expert, labels *Roe v. Wade* "a very bad decision." It is not bad, he writes, solely because abortion is an issue better left to the legislative process. "It is bad because it is bad constitutional law, or rather because it is *not* constitutional law and gives almost no sense of an obligation to try to be."[25] Ely is especially critical of the Court's "rush toward broader ground" to identify a general right of privacy. If, by "privacy," we mean freedom to live one's life without governmental interference, the Court did not even imply the term, for, as Ely notes, such a right is at stake in *every* Supreme Court case.

Even *Griswold* had more rational grounds, the Court invalidating only that part of the Connecticut statute that forbade the *use*, as opposed to the manufacture, sale, or other distribution, of contraceptives. In another right of privacy case, *Katz v. United States* (1967), which limited governmental tapping of telephones, the Court interpreted what the framers might reasonably have called a "search." Therefore, whereas it is entirely within the Court's domain to identify specific zones of personal privacy, or even to imply a general right, Ely insists that some care should be taken in defining "the sort of right the inference will support." In the end, the Court would have been more open and honest had it endorsed abortion rather than privacy. It may not have quieted the public debate, but it surely would have lessened the confusion over privacy.

In the selected cases discussed, the Court over a span of time appears to have been consistent, albeit a paradoxical consistency, in wanting the best of both worlds, the private and the public. On the one hand, the justices have developed some notions about the "fundamentalness" of personal and private activity; but, on the other, they seem to endorse a "general" right, albeit vague, to leave the door open for the protection of future fundamental activity. In *Union Pacific Railway Co.*, which restrained a court's authority to order the revelation of private information in a civil suit, the Supreme Court favored the basic right of self-control over civil intervention. In *Boyd*, the sanctity of the home was at issue. Other cases dealt with implied but specific, if not always fundamental, private activities. From these it is possible to decipher which rights the Supreme Court has deemed "fundamental;" it is not as easy to discern the "general" right, however.

Because the framers of the Constitution deemed political rights as the most fundamental, indeed, the very foundation of the Bill of Rights, Americans no longer debate the significance of free speech, press, assembly, their right to petition government, and a broad voting franchise. But when we decide to recognize some privacies of life and ignore others, we decide to recognize some general right to make personal choices. As Paul Bender sees the problem: "If the Constitution is going to protect an area of private choice for *private* reasons, we cannot judge the fundamentalness of certain choices—as opposed to others—through their external appearance or effect."[26]

The right to enjoy sex in marriage, for instance, cannot be more fundamentally private than the right to enjoy sex outside of marriage, because, as Bender posits the paradox, that would reflect a choice that marriage is a better state than nonmarriage, and that is "precisely the decision which a right to privacy must leave to the individual, if marriage is, indeed a private matter." And, because it is impossible to derive a test for determining "fundamentalness," the only way to make the critical determination is to identify the private nature of the activity. Moreover, if it is truly private—that is, "fundamental"—then there is no "compelling" reason to regulate it. Societal interest, clearly, must be a determining factor, not the private pursuit of private interests or private activities.

With such an approach, the Court's role is to serve as arbiter between the strengths of the private versus those of the public, those of the individual as opposed to those of society. Occasionally, as we have seen, the private interests must prevail, but usually they need not, because most private interests can be satisfied without harming or implicating others. As Bender notes:

The Court should not go too wrong in this area if it acts in the name of privacy only when it perceives the grossest kind of disparity between the detriment to the private pursuit of happiness which a law causes and the happiness the law is meant to protect.

The right to privacy as the right to be let alone has come a long way since Warren and Brandeis sought protection against gossip in the press. In addition to contraceptives and abortion, education is a private activity, travel is a private activity, long hair is a private activity, as are a number of other activities of the individual that American courts seem happy to accommodate under the aegis of "privacy." Yet, this tendency, especially pervasive in recent Supreme Court rulings, to equate personal privacy with individual freedom, or autonomy, is neither surprising nor out of line with the current public yearning for basic human rights.

Thinking on the question has shifted, from concern for the aggregate of individuals to concern for the separateness of human beings, and may be attributed to the move away from the utilitarian concept of public welfare as a viable societal goal. There is now a revival of the doctrine of basic human rights, protecting specific liberties and interests of the individual. Utilitarianism has been under the gun, in fact, at least since Jeremy Bentham proclaimed nearly two hundred years ago that government and the limits of government were to be justified by the greatest happiness of the greatest number, and not by any doctrine of natural rights, which Bentham thought so much *"bawling* upon paper."

Because utilitarianism may still provide valuable insight, we now turn to that philosophy to establish the rationale for privacy in the context of what is more important, publicity.

5

Rationale of Public

With this chapter and the next, the halfway point has been reached. Privacy as a historical reality and a legal entity was treated in the first four chapters; perhaps not definitively but sufficiently to serve as a basis for the final four, which deal with contemporary problems surrounding the ongoing search for privacy. So far, however, certain assumptions have gone unchallenged. The first is that privacy is a human right; the second is that this right is a natural one, as if preordained and transcendental. This chapter and the next are meant to establish privacy as a right, but a utility right rather than a natural right, and to present an alternative way of thinking that both protects individual solitude and enhances the public weal. In this chapter, a case is made in support of public openness based upon democratic principles, and in the following, arguments are presented on behalf of a sense of community based upon persistent commonality among individuals. Because both chapters explore philosophical issues, Jeremy Bentham, the great utilitarian theorist, and his descendants and detractors can usefully be given center stage.

> Men are born free and remain free, and equal in respect of rights. Social distinctions cannot be founded, but upon common utility.

This is Article I of the 1789 French "Declaration of the Rights of Man and of the Citizen," with which Bentham was at first sympathetic, but within two years he began to be disillusioned by the radical turn the French Revolution was taking. He had been fearful of the anarchic implications of the American Declaration of Independence in 1776 but was willing to tolerate the French Declaration, until citizens were put to death and property was threatened for reasons he believed to be foolish. Thus, by 1795 he had begun to compose a work to be called "Pestulance Unmasked!," which was published a year later as *Anarchical Fallacies*, his condemnation of the French Declaration and the actions perpetrated in its name.[1]

"All men, on the contrary, are born in subjection, and the most absolute subjection," Bentham wrote in response to Article I. "In this subjection he continues for years . . . and the existence of the individual and of the species depends upon his so doing."

> The end in view of every political association is the preservation of the natural and imprescriptible rights of man. These rights are liberty, property, security, and resistance to oppression.

Bentham's response to Article II is a summary of his lifelong opposition to natural rights.

> More confusion—more nonsense—and the nonsense, as usual, dangerous nonsense. For natural, as applied to rights, if it mean anything, is meant to stand in opposition to *legal*. In proportion to the want of happiness resulting from the want of rights, a reason exists for wishing that there were such things as rights. But reasons for wishing there were such things as rights, are not rights; a reason for wishing that a certain right were established, is not that right—want is not supply—hunger is not bread. *Natural rights* is simple nonsense: natural and imprescriptible rights, rhetorical nonsense, nonsense upon stilts.[2]

Bentham continued in this intemperate vein, attacking the language of rights he believed the French revolutionaries ("so many terrorists," he said) had severely abused and, in so doing, had raised a number of anarchical fallacies, or, simply, false expectations. He abhorred revolutionary measures, and had seen in his lifetime two major revolutions defended by such declarations. In the case of the American document, Bentham eventually came to embrace the resulting United States—although he continued to believe that the country had been founded on bad arguments. But in the French, he found reason to regard natural rights as the language of anarchists and terrorists. "Is not the liberty of doing mischief liberty?" Bentham asked.

Concluding his analysis of the French declaration, Bentham returned to his own notion of right, or rights:

> *Right,* the substantive *right* is the child of law; from *real* laws come *real* rights; but from *imaginary* laws, from laws of nature, fancied and invented by poets, rhetoricians, and dealers in moral and intellectual poisons, come *imaginary* rights, a bastard brood of monsters, 'gorgons and chimeras dire.' And thus it is, that from *legal rights,* the offspring of law, and friends of peace, come *anti-legal rights,* the mortal enemies of law, the subverters of government, and the assassins of security.

Or, as Mary P. Mack, a biographer, has paraphrased his position: "If a man had a right to something, all it could mean was that government would guarantee his use and enjoyment of it. Rights were legally guaranteed patterns of behavior."[3] L. W. Sumner adds this interpretation: "The naturalness of natural laws seems to exclude their being laws, and thus the naturalness of natural rights seems to exclude their being rights."[4]

Utility, Bentham's basic philosophic and legalistic construct, finds expression in a unified theory of law, whose basis is a strong state characterized by ethical individualism. But natural rights—"every villain's armoury, every spendthrift's treasury"—fight against the effort to systematize law, which Bentham insisted was better achieved through legislation. This kind of law best serves the individual too, for, as Nancy L. Rosenblum interprets Bentham, "The state can protect men better than any other form of order."[5] Rights which are legislated, rather than "found" in some natural law or "discovered" by a panel of judges, are more likely to reconcile men to diversity and change, because the legislative process implies consent of the governed. For instance, Bentham said that the "right to property" must be precise, perhaps modified, even limited, before it can help to secure one's possessions. The "right to privacy," historically tied to property, must also be dealt with by reasonable and consensual agreement.

Bentham spoke of the state as "political society" to distinguish it from government as an all-powerful separate entity and also to emphasize the state's fundamental commitment to the individual. Because much of what Bentham wrote was addressed to political leaders, especially legislators who had a penchant for secrecy, he insisted that governments, as utilitarian bodies, exist only at the behest of citizens and their happiness. Individualism is not lost but is a correlative of obedience, which in turn is the essence of popular sovereignty for the state as well as for the individual citizen. "The idea of *law, offence, right, obligation, service* . . . are born together . . . exist together, and . . . are inseparably connected." Individual rights, central to Bentham's theory of representative government, are bestowed rather than natural. "Whatever is given for law by the person or persons recognized as possessing the power of making laws, is *law*."

What is usually implied by those modern writers on privacy who criticize the collectivism of utility theory is the naturalness of privacy—from Warren and Brandeis to Bloustein and Westin. Yet human rights built on apt slogans are as likely to be destructive of society as they are to serve the human need for solitude and seclusion. Because Bentham consistently preached the utility of a strong state, he would have endorsed one of the concerns of this volume: that privacy affords an escape from the obligations and burdens of public life and threatens collective survival. If the state is to survive, the individual must be more public than private.

Utilitarianism, as developed by Bentham and his followers, is not without problems, however, the obvious one being its "will of the majority" overriding and perhaps suppressing minority opinion, as John Stuart Mill noted in his famous critique in 1838. Bentham had sought to serve the rising middle class through reforms directed against the privileges of the aristocracy. When Mills's essay *On Liberty* appeared in 1859, the central concern was with the preservation of the rights of individuals and minorities against public opinion and the democratic state. Where Bentham had asserted that good government depended upon the moral responsibility of the governors, as well as the citizens' obedience to the rule of law, Mill focused attention on individuality, but individual action limited by the rights of others. Society has a right to protect its members from "moral vices," so that law supplants liberty when private action risks damage to others.[6]

Both men opened up society to all individuals and made individuality a counterbalance to majoritarian tyranny. They both, but especially Bentham, emphasized legislated law over natural or moral law. For, as A. C. Dicey wrote at the turn of the century,

> Legislation deals with numbers and with whole classes of men; morality deals with individuals. To ensure the happiness of a single man or woman . . . is a task impossible of achievement. To determine, on the other hand, the general conditions which conduce to the prosperity of the millions who make up a State is a comparatively simple matter.[7]

Dicey, a professor of law at Oxford, also noted that law is concerned primarily with external actions and only indirectly concerned with motives. Morality, however, is concerned primarily with motives and feelings—only indirectly with actions. Thus, Dicey said, the principle of utility is a good test of the character of a law. Because law in our time has become so central to whatever it is that society tries to accomplish, how we perceive the role of law, in addition to the rights we attach to law, becomes of primary importance.

Iredell Jenkins, writing in 1980, elucidates three cardinal

points that governed Bentham and his followers and that apply to our own wish for legislation that is indeed utilitarian. First, Bentham recognized that law and social reform cannot provide happiness, but can only promote the conditions that will enable citizens to procure happiness for themselves. Applied to privacy, this means that not all elements of the private life may need to be protected, at least not without the benefit of reason and utility. Second, Bentham believed that reforms can succeed only if they are appropriate to the circumstances of society, "consonant with the feelings, habits, and desires of men, and commensurate with the material resources available."[8] Applied to privacy, this means that perceptions and yearnings need to be tempered (albeit delicately) by reality. Richard Rovere, the late American political journalist, addressed this point when he wrote, "We were willed a social order dedicated to the sovereignty of the individual but, thanks mainly to technology, dependent for its functioning largely on the interdependence of lives."[9]

Bentham insisted as the third cardinal point that no reform measures should be undertaken, or even proposed, without assurance that the legal apparatus had sufficient command of the necessary resources—physical, human, and institutional—"to give a high probability that the ends in view could actually be achieved." This suggests that, in the case of privacy, we have relied too heavily upon the courts instead of appropriate legislatures for the enhancement and protection of individual rights. Representative government is simply a better forum for the determination of utility law. It was argued in an earlier chapter that the Supreme Court on several occasions has defied this probability clause of modern self-governance. By expanding its "zones of privacy" to include nearly all that is nonpublic in personal life, the Court has diminished the real value of the individual. It has left little to private choice, which is the kind of individuality envisioned by utility theory.

"If a person possesses any tolerable amount of common sense and experience," Mill wrote, "his own mode of laying out his existence is the best."[10] Government interference was not advocated by either Bentham or Mill, except for a greater good; instead, they urged individuals to assume responsibility for deriving benefits that the state is incapable of providing or should not attempt to provide. If privacy is so strong a desire as to be beyond the individual or the collective citizenry to protect, then and only then should government, and the courts in particular, intrude.

The basic danger in treating privacy as an inalienable human right, as opposed to a utility right, is twofold: first, as Bentham reasoned, there is the problem of applying man-made law to something so general and abstract as given, if not solely possessed, by God; and, second, as utility theorists generally have posited, there is the problem of justification—moral, political, or legal—because definitions of privacy change. The human rights of personhood, autonomy, liberty, equality, and even happiness, have always been valued highly, usually as absolutes. But the ground underneath shifts, conditions alter, and the sociology of one age makes that of another irrelevant. Values, too, may change, but even if they remained constant, rights based upon utility have a better chance of survival than those based upon principle alone. Edmund Burke, in his own reflection on the French Revolution, wrote:

> What is the use of discussing a man's abstract right to food or medicine? The question is upon the method of procuring and administering them . . . Rights undergo so many refractions and reflections it is absurd to talk of them as if they continued in the simplicity of their original direction.[11]

Utilitarians find most rights-based theories circular in concept and impractical in application, high-minded but not very realistic. The demand for rights is inferred, not from the utility of rights, but from some basic principle such as liberty or equality. Jeremy Waldron, a political theorist, attacks such thinking as contradictory and self-seeking: "The ambiguity of the concept enables rights to feature both as *explanans* and *explanandum*, and the philosopher never has to break out of the charmed circle and confront any of the hard questions of real life.''[12] The point has never been put better than by Bentham himself:

> The dictates of reason and utility are the result of circumstances which require genius to discover, strength of mind to weigh, and patience to investigate. The language of natural rights requires nothing but a hard front, a hard heart, and an unblushing countenance. It is from beginning to end so much flat assertion. It lays down as a fundamental and inviolable principle whatever is in dispute.[13]

It is interesting to note that Great Britain has never recognized invasion of privacy as a legal matter. A Parliament-created committee on privacy in 1972 recommended against such a move because it would be extremely difficult to delineate a right of privacy, and even so the threat to free speech and free press would be at too great a risk. Utilitarian philosophy may have influenced the committee in its report, for Bentham once warned:

> To know whether it would be more for the advantage of society that this or that right should be maintained or abolished . . . the right itself must be specifically described, not jumbled with an undistinguishable heap of others, under any such vague general terms as property, liberty, and the like.[14]

Unless rights are specified and utilitarian, they are nothing but "rhetorical nonsense" and "*bawling* on paper."

Iredell Jenkins picks up on this important point in her thoughtful study, *Social Order and the Limits of Law* (1980), in which she says that because rights are usually rooted in some ultimate metaphysical ground, any emerging doctrine of rights quickly becomes a powerful rhetorical tool. We clothe our desires in the language of rights. Legislative acts and judicial decisions are based on ethical values and citizens are expected to honor the rights of others as a moral duty. Privacy is accorded natural *and* legal status, the latter by virtue of statutes and rulings, which, in the twentieth century, are designed to enforce respect for human rights. Whether legal or meta-legal, rights are either conceived and created by law, "or they exist as aspects of reality prior to their legal annunciation and are merely recognized by law.[15]

Jenkins identifies two essential characteristics of rights: obligatoriness and instrumentality (deontological and pragmatic). They are obligatory because they are declared to be ones that "ought" to be recognized and respected. Rights are instrumental in that they function as means to ends, hence the primary locus of rights lies in the perception of injustice. Rights, according to Jenkins, have their origin in the effort to correct a wrong, an undeserved harm. Thus conceived, rights are free of content; they seem to need only to be proclaimed. Jenkins applies both rights-based theory and utilitiarianism when she asserts that we can make distinctions between rights and simple desires by claiming as rights those things due us upon consideration of the nature of the world and society. "Rights may have to be won, but they do not have to be earned or merited."[16]

Although there is no question that rights per se deserve some official sanction and protection, there remains the yet unfulfilled need for some objective clarification of both the ground and the content of specific rights. Jenkins's conclusion is poignant: "Natural rights were . . . conceived as functional rather than substantive; they were designed to afford opportunities, not to assure outcomes."[17] It follows, therefore, that a natural "right" to privacy is simply inconceivable as a legal right—sanctioned perhaps by society but clearly not enforceable by government. The law may state that others are prohibited from intruding upon one's solitude and seclusion, one's natural affinity for privacy, but it could not guarantee the need for autonomous behavior in the first place. Even if privacy were determined a universal human right, no law, or laws, could protect that right beyond establishing a priori a mechanism for detecting inappropriate, or illegal, invasions. Privacy itself is beyond the scope of law. The accuracy of this statement was apparent in earlier chapters on the development of conflicting judicial efforts to constitutionalize privacy. It is also evident in later chapters on the failed efforts to legislate privacy and freedom of information at the same time.

What is required is a process for transforming natural rights, which are embedded in societal morals and various class and group mores, into legal rights, which, at least since Bentham, have demanded governmental attention. First, the process must refine the very concept of legal rights and acquire a catalogue of such rights. Second, the process must develop a substantive procedural mechanism for protecting the exercise of rights and remedying their violation. "Natural rights are merely claims, regardless of the intellectual justification and emotional fervor with which they are pressed," Jenkins maintains. "Legal rights give title, backed by force."[18] It is difficult to predict how many rights can be legalized under this kind of scrutiny, for the "rights" to equality, liberty, property, reputation, happiness, and privacy are acknowledged in varying degrees in our culture and subcultures, but it is hard to reduce them to common definitions. The legalization process is largely isolated from the needs that initiated it and the values that it envisaged, Jenkins points out. Most legal rights that once were natural phenomena have evolved haphazardly, unconsciously, and, in the case of privacy, uncritically.

"Human rights," the third set of rights categorized by Jenkins, are not unlike natural rights in that they are born of desperation and dedication. They resulted in the American Declaration of Independence, the French Declaration of Rights, and, more recently, the United Nation's Universal Declaration of Human Rights, each of which followed a difficult period in history. Human rights (sometimes called "civil rights") appeal to feelings and emotions, seldom to the intellect. But, like the appeal of natural rights, they serve to revive an interest in rights in general and help to ensure the relevance of the individual to society at large. "If rights are human, then man's their measure."[19]

No limits are placed on what may be *claimed* as a human right. Today there are rights claims to low-cost housing, a food allowance, medical assistance, education, a minimum wage, and amnesty. These are *due* people as human beings and so are *owed* them by society. Yet, as Jenkins properly asserts, the concept of natural rights teaches us that any doctrine of rights must have a firm theoretical basis. Furthermore, the doctrine of legal rights teaches us that declarations of rights are meaningless without an effective apparatus to implement them.

"Privacy is desirable, but rights to enjoy it are not absolute," writes the law professor Hyman Gross. When the U.S. Supreme Court ruled birth control a privacy right in its *Griswold* decision, it incorrectly equated privacy with personal autonomy and, so doing, stymied any serious efforts at forming a theoretical base for privacy as a legal right. The justices saw "right of privacy" and "right to be let alone" as synonymous. "In the *Griswold* situation there had been an attempt by government to regulate personal affairs," Gross writes, "not get acquainted with them, and so there was an issue regarding autonomy and not privacy." Whereas an offense to privacy is nearly always an offense to autonomy, not every curtailment of autonomy is a compromise of privacy.[20] This is not to say that the two concepts— privacy and autonomy—are unrelated; rather, it is to suggest that the two enjoy a paradoxical intimacy. As Ruth Gavison points out in an essay on the availability and exclusion of private information, it might be argued that privacy will work *against* autonomy "by encouraging a person not to stand up for his commitments," which is a desirable democratic and public act. She notes that, while privacy encourages intimate relationships and the confidentiality of personal information, the growth of a relationship based upon trust may also require its public acknowledgment. "Marriage owes much of its symbolic importance to its public nature."[21]

These are shades of Jeremy Bentham, whose rationale of public openness called for realizing the limitations of imprescriptible rights, even those of liberty, property, security, and resistance to oppression, the guarantees for which resided with responsible and responsive governors. A final observation by Iredell Jenkins is at once appropriate and persuasive:

Traditional legal rights are primarily protective: they guarantee citizens certain basic freedoms and immunities, and they protect them against intrusion or arbitrary action by the state. These rights do not bestow any positive benefits upon people. They do not promise to provide arms, an enlightened press, churches, places to assemble, legal fees, or houses and other property: they merely assure the citizen that the state will not interfere with his pursuit of these goods nor deprive him of what he has won by his efforts. If traditional rights are to bear fruit, they must be exercised by those who hold them, since by themselves they merely give assurance of an open field and fair play (at least within the rules of the game).[22]

It should be apparent by now that privacy is best conceived of and dealt with as a utility right, based upon natural desire or right, but really only viable if made a legal right with other societal needs and rights in mind. Because privacy is so nebulous in concept, it is well-suited for utilitarian resolution. To the glib charge that utility theory does not take seriously the distinction between individuals, James Griffin insists that it is a mistake to think our moral "intuitions" are always better engaged by such slogans than by the principle of utility. Thus, he says, "there is nothing in the formal conception of utility that rules out one value's being incommensurable with another," but it may be necessary on occasion to sacrifice some rights so that a greater number of others may be upheld.[23]

Privacy, like its sibling, autonomy, is a matter of sovereignty "at the center of one's life," but, as it relates to "the personal commitments central to one's life plan," it is not a very extensive right. What, for instance, does it tell us about the invasions of privacy that worry us now: telephone-tapping, electronic eavesdropping, access to one's medical or financial records? Why do governments want to tap telephones? To these Griffin adds the most significant query: "Which is now the greater danger— public intrusion in private life or private subversion of public life?"

Privacy today threatens not only a traditional American commitment to public service, but it endangers another proud trait—individualism. "Our individualism," writes Michael Kammen, ". . . has been of a particular sort, a collective individualism. Individuality is not synonymous in the United States with singularity. When Americans develop an oddity they make a fad of it so that they may be comfortable among familiar oddities." Their unity, as Ralph Waldo Emerson wrote, "is only perfect when all the uniters are isolated."[24] Utility theory and individualism are indeed compatible. Utilitarians such as Griffin merely argue on behalf of some rational means to deal with justifiable human rights, privacy among the important ones. His thinking helps point the way:

> What is at stake in debates about modern intrusions on privacy is a certain sort of power, and if nowadays the greater threat is from government power, and if the threat is great enough to undermine one's life plan, protection should go to individuals. But this sort of argument is not timelessly valid; it allows the possibility that at other times, in other circumstances, the powers should go elsewhere.

Rights are thus subject to periodic redrawing, determined by practical considerations. And because pragmatism enters into the determination of virtually every human right, "human rights have neither sharp nor fixed edges."[25]

Classical utilitarianism, as delineated by Bentham, maintained that privacy was the opposite of publicity, or openness, and identical to secrecy, described as one of the most "mischievously efficient instruments of despotism." Publicity was among the natural instruments of justice, but Bentham advised that "on no occasion to give to privacy any extent beyond what the particular nature of the occasion absolutely requires. Justice seems to have everything to gain, nothing to lose"[26] by employing privacy sparingly. He saw publicity and privacy as opposite and antagonizing but mutually connected qualities. He believed that the highest conceivable degree of publicity can do no harm, at least to the purposes and ends of justice. Privacy should generally "wear the character of *exceptions*."[27] The openness generated by publicity operated as a check upon mendacity and incorrectness. And, in keeping with Iredell Jenkins's desire for a theoretical basis for human rights and an apparatus for implementing them, Bentham advised: "Good *laws* are such laws for which good reasons can be given; good *decisions* are such decisions for which good reasons can be given."[28]

Bentham allowed, however, that privacy was necessary on occasion, as, for instance, the prevention of needless violation to the reputation of individuals and the peace of families. "The worst mischief arising from publicity is always a limited one, whereas the mischief attached to inviolable secrecy is altogether boundless."[29] Arbitrary power is the result of perpetual and indefinite concealment. "Let no degree of privacy be produced in waste," advised Bentham. "The protection of the public eye is a security too important to be foregone."[30] And, as if to address the Fourth Estate:

> Under a mixed constitution like the British or a republican constitution like the Anglo-American, the difficulty would be to prevail on the people to view with complacency any such extent given to the principle of privacy as the dictates of abstract utility might be thought to require.[31]

Bentham's position has not gone without its detractors. Robert Nisbet, the twentieth-century political philosopher, calls for a revival of "the prestige of the private, as contrasted with the public." If there is to be "an efflorescence of a truly free and also stable society," Nisbet seeks a reversal of the tendency since the nineteenth century to make the public ethically superior to the private. Rousseau, with his doctrine of the General Will, is to blame, then Bentham, "with his hatred of all traditional privacies, his veneration of the collective." They perpetuated the conviction—quite rightly, one might add—that what is public is inherently better than what is private. Besides being an overreaction to (and a misreading of) Bentham's view of the private life, Nisbet's analysis seems to throw out the baby with the bath water. For at the heart of Bentham's notion of utility is the private person who is indeed subordinate to the collective, but because it is through such strength in numbers that the individual gains freedom and society enjoys stability.

"It is, as we know, difficult for the modern liberal to believe that the two kinds of invasion of personal privacy are closely related," Nisbet writes, "that invasion which begins in the name of Plato or Rousseau commonly winds up as invasion in the name of CIA and FBI."[32] Taken seriously, the assertion lacks congruence with even historiography's invisible hand theory; taken in the vein it must have been written, it assumes a rationale that even KGB operatives are not likely to invoke before surveillance. The worry is real, however, although it is inaccurate to think of Bentham or his followers advocating wiretapping, bugging, and covert military activity in the name of justice. Bentham advocated openness precisely because he believed that corruption and the abuse of freedom were inherent in governmental arrangements. What indeed is difficult for present-day liberals (some of whom are now conservatives) is utilitarianism's juxtaposing of morals and legislation, the art and science of which Bentham said depended upon enlightening the people to the elements of pain and pleasure. And he said it was the task of the few to enlighten the many. Bentham's optimism and patience were incomparable, for, as James Steintrager interprets his philosophy, "The feelings of the people would have to be changed by persuasion and not by force."[33]

Rather than blame Bentham and the other nineteenth-century reformers for today's misadventures, it would be wise to look to the nature of a society that obscures values and resists a kind of public dimension for private freedom. One must wonder whether freedom to choose without giving serious consideration to the effects upon others is a freedom in the interest of society at large, or if, indeed, it is a freedom at all. What may be at the root of the genuine concern Nisbet and others voice is not privacy's dwindling prestige, but, conversely, a decline in sympathy for the collective, the similarity among individuals in a community, a diminution in commitment to public life. Only a pessimist would find solace in the metaphor that life is simply a matter of rearranging deck chairs on the *Titanic,* but an optimist might encourage passengers to care more about their intended destination. (A pessimistic opportunist might insist on "first-class" accommodations as well!) Steintrager concluded his thoughtful study of Bentham with this typical utilitarian challenge:

> One may well prefer to begin to re-examine the nature and purpose of political thought by confronting the problems left unresolved by Bentham's endeavours, rather than accept the all too easy solutions of many who came after him.[34]

Somewhat utilitarian, although not strictly, is the work of the contemporary legal theorist Ronald Dworkin, whose central thesis in essays and books is a preference for equality rather than liberty, or *in addition to* the freedoms that liberty entails. Dworkin essentially is a rights-based theorist, who, unlike Bentham, seeks to defend individual claims against general societal interests. Although he does not fully embrace the "greatest happiness of the greatest number" as the primary goal of government action, Dworkin seeks a society that is "better off in an ideal sense." A more equal society is a better society, thus a community that is "more equal" is therefore "more just." Because Dworkin's basic thesis suggests a *claim* to equality, it can be said that his notion of rights is somewhere between the legal and the natural, or meta-legal, presupposing equality as a value for society.[35]

Prominent in Dworkin's schemata is the distinction between the argument of *policy* and the argument of *principle*, which is yet another way of parsing the rights rhetoric, but, in keeping with utility theory, laced with a strong sense of reality. His policy argument is based upon furthering and protecting some collective social goal, where principle is grounded on an appeal to the individual or some social right. "Hard cases," those which cannot be brought under a clear rule of law that is already established and in which judges have some discretion, should be generated by principle not policy.[36]

Such a case was that of a *New York Times* reporter who was imprisoned and his newspaper fined for failure to turn over confidential material to the judge in a murder trial. The newspaper and its reporter, Myron Farber, argued that the judge had acted illegally on two different grounds: the order violated New Jersey's "Shield Law," which stipulates that in any legal proceeding a newsman has a privilege to refuse to disclose any source of news or information obtained "in the course of pursuing his professional activity"; also, that the judge's order violated their rights under the First Amendment, which provides for "freedom of the press." Those, who, like the judge, believe that the rights of free speech and publication, although of fundamental importance, are not absolute and must sometimes yield to competing rights, are employing Dworkin's argument of principle. There is some congruence here with utility theory, for the reporter's individual rights are subsumed by society's collective interest in justice. But, as Dworkin explains in his own words below, the principle-policy nexus is not always supportive of individuals as individuals, but rather ensures equality *among* individuals under a given code of law. Dworkin explains the difference:

> Justifications of *principle* argue that a particular rule is necessary in order to protect an individual right that some person (or perhaps group) has against other people, or against the society or government as a whole. Antidiscrimination laws, like the laws that prohibit prejudice in employment or housing, can be justified on arguments of principle.
>
> Justifications of *policy*, on the other hand, argue that a particular rule is desirable because that rule will work in the general interest, that is, for the benefit of the society as a whole. Government subsidies to certain farmers, for example, may be justified, not on the ground that these farmers have any right to special treatment, but because it is thought that giving subsidies to them will improve the economic welfare of the community as a whole.[37]

Sometimes, of course, principle and policy argue in opposite directions, and when they do, Dworkin believes, policy must yield to principle. Here is where Dworkin parts company with utilitarians, for, although fluidity in policy decisions appears to allow for a collective good, decisions of principle appear to adhere to some rigid concept, a natural right, if you will. In the Farber case, the principles involved were those of a defendant's right to a fair trial and the newspaper's right to free speech, neither of which had been violated by the judge's order. Newsmen do not, as a matter of principle, have any greater right of free speech than anyone else, despite the fact that, as a matter of policy, their special privilege may result in improved education among the citizenry as a whole. But it's the principle of the matter that is more important to Dworkin because principled decisions ensure equality. The applied logic is fascinating, although Dworkin walks a fine line between equality in theory and equality in practice as his reasoning becomes more specific:

> If free speech is justified on ground of policy, then it is plausible that journalists should be given special privileges and powers not available to ordinary citizens, because they have a special and indeed indispensable function in providing information to the public at large. But if free speech is justified on principle, then it would be outrageous to suppose that journalists should have special protection not available to others, because that would claim that they are, as individuals, more important or worthier of more concern than others.[38]

What this teaches is that the journalist's right does not translate into a societal benefit, but that a society's right becomes a benefit for journalists as well. What strikes the writer as ironic is that by appealing to a principled application of law, Dworkin is endorsing an open equalitarian society. If only certain people had rights, the very idea of rights as a universal concept, meta-legal or legal, would hold little real meaning. Ironic, as a way of neatly describing Dworkin's position, may be a misuse of the term, but it applies nonetheless because the real equality of law most certainly depends upon equal resources and equal welfare among the citizens upon whom otherwise principled law is applied. That is, if members of a society are not equal in real and genuine terms, then the application of either principle or policy lacks any justification whatsoever. Even among journalists, who pride themselves on being equal, there is inequality of constitutional protection because of the unequal distribution of resources, largely financial but often legal, to enable them to enjoy equality under the law. Dworkin's scheme is philosophic but hardly realistic. James Griffin suggests an alternative worthy of the philosopher's time:

> Every plausible principle of equality is based on the thought that everyone matters and matters equally, and to stress only formal features of distribution is to remember the *equally* but to forget the *matters*. It is not just one's capacity to live out a life plan that is valuable; the quality of life that one manages then to live also matters.[39]

Despite the convincing nature of utility theory, the swing in moral philosophy in recent years has been in the direction of a more conservative approach to human rights. By and large, rights theorists base their position on the moral importance of the separateness and distinctness of persons, which traditional utilitarianism is said to have ignored. Examples are the work of Robert Nozick and Dworkin, who have little else in common except their commitment to individual rights. John Rawls, too, is an eloquent spokesman for rights, and, like Dworkin, laces his conservatism with equal regard for the rights of others. But it is Nozick whose work heralds the swing to the right.

Nozick advocates a strictly limited set of near absolute individual rights, which express "the inviolability of persons" and reflect "the fact of our separate existences." Each individual, so long as he does not violate the same rights of others, has the right to be free from all forms of coercion or limitation of freedom. Nozick's moral landscape, to borrow from the British legalist H. L. A. Hart, contains only rights and is empty of everything else except possibly the moral permissibility of avoiding what he terms "catastrophe."[40]

For Nozick moral wrongdoing is simply the violation of rights, perpetrating a wrong to the holder of a right. What distinguishes this view from that of Dworkin or Rawls is that it matters little how a social system actually functions, "so long as rights are not violated," even if the system produces misery or inequality. In particular, the state may not impose burdens upon the wealth or income, or restraints upon the liberty of some citizens in order to relieve the needs or suffering, however great, of others. Nozick believes that the best state is the one that protects individuals, that it may not go beyond the night-watchman functions of protection against force, fraud, and theft or breaches of contract. The violation of individual rights is the one, and perhaps only, form of moral wrong.

Nozick's philosophy appears to disregard any harm that individual rights may impose on society at large, and for him to imply otherwise would of course undermine the separateness of individuals. Although Dworkin is not in the strict sense a classical utilitarian—it is wrong for government to deny a right once it has been defined—he nevertheless disputes the existence of a general or residual right to liberty, as is central for Nozick. Hart's analysis is cogent:

> There are only rights to specific liberties such as freedom of speech, worship, association, personal and sexual relationships. Since there is no general right to liberty, there is no general conflict between liberty and equality, though the reconciliation of these two values is generally regarded as the main problem of liberalism; nor, since there is no general right to liberty, is there any inconsistency, as Conservatives often claim, in the liberal's willingness to accept restriction on economic but not on personal freedom.[41]

General interest and general welfare are at the center of Dworkin's neo-utilitarian philosophy. (Although he may deny affiliation with, or indebtedness to, Bentham, Dworkin's argument, like Bentham's, has a certain "Byzantine complexity," according to Hart.) Dworkin departs from doctrinaire utility theory in his commitment to certain liberties so important that they ought never to be overridden, even in order to advance the general welfare. The success of his approach to rights depends a great deal, perhaps too much, on an altruistic society that is willing to honor general rights without the benefit, or force, of law. Majoritarian rule would protect the absolute rights of the minority as well.

Nozick's effort to derive rights from the seemingly uncontroversial notion that individuals are separate and distinct is fetching in its apparent simplicity. Dworkin's attempt to assign rights on the basis of equality is equally as intriguing but far more difficult to adopt. As discussed earlier, Dworkin wants to separate the principle of rights from the policy of rights, the former to be used as a guide for society's determination of the latter. Both philosophers are working in the shadow of utilitarianism, but neither of their approaches to rights is likely to provide a realistic context for dealing with privacy.

Because Nozick is concerned with the individuality of persons, he supports the fundamentalness of the right to personal privacy. Dworkin, too, supports a principled privacy, with specific rights to be limited and those less specific to be identified on a case-by-case basis. Nozick is likely to engender more of what Bentham called "anarchical fallacies," and Dworkin, an egalitarian hodgepodge. Instead of basing future judgments regarding privacy as a human right on "separate existences" or "equal concern and respect," the belief here expressed is that utility theory combined with a genuine sense of community could honor and protect both the individual and the society.

If it is possible to set aside the factor of government for the sake of instructive debate on the more fundamental issues, then it is within the range of probability to have it both ways—to conserve individual freedom at the same time a liberal society enforces the rights of all its citizens, if not by law, then by custom. In this, a utilitarian community, instead of an authoritarian regime, becomes *the* major factor for accommodating privacy in a public setting.

6

A Sense of Community

Do I contradict myself?
Very well then I contradict myself,
(I am large, I contain multitudes.)

WHITMAN, *Leaves of Grass*[1]

Americans have always had it both ways. Erik Erikson, the distinguished psychoanalyst, said that any identifiable truly American trait can be shown to have its equally characteristic opposite: "This dynamic country subjects its inhabitants to more extreme contrasts and abrupt changes during a lifetime or a generation than is normally the case with other great nations." Most Americans, Erikson maintained, have been confronted with a series of alternatives, or "opposite potentialities"—open roads of innovation and jealous islands of tradition, outgoing internationalism and defiant isolationism, boisterous competition and self-effacing cooperation. Thus, the functioning American, as the heir of a history of extreme contrasts and abrupt changes, according to Erikson, bases his final ego identity on some tentative combination of dynamic polarities, such as migratory and sedentary, individualistic and standardized, competitive and cooperative, pious and free-thinking, responsible and cynical.[2] To these may be added the polarities of personal privacy and public commitment, individualism and community.

In his examination of the antebellum years, 1836–1860, E. Douglas Branch touched on the same theme with a certain comic irony:

> Beyond the area of middle-class dominance lay the frontier, a proud, crude society rampant in contradictions—democratic

equalitarianism and individual license, laissez-faire and pater-
nalism, malaria and insatiable vigor, quinine and corn whiskey.[3]

America's nineteenth-century seer Ralph Waldo Emerson once
insisted about his countrymen, "We want our Dreams and our
Mathematics." And Michael Kammen, in our time, observes
that the "quintessential American hero wears both a halo and
horns."[4] The late Warren I. Susman, writing of the 1930s, notes
that, for a culture that originally had enshrined individualism as
its key virtue, interest in the "average" was now overwhelming:
"The Average American and the Average American Family became central to the new vision of a future culture."[5] Perhaps the
term that most aptly describes the American way is Kammen's
"contrapuntal civilization," the origins of which he traces so
poignantly in his *People of Paradox.*

Kammen points out that the United States is commonly represented by two images, or icons, both of which appear on posters
and in cartoons, but it is never really clear which is most appropriate at any given time or circumstance. One is the tall, thin,
goateed old man attired in a formal suit cut from a Betsy Ross
creation. The other, Kammen describes, is a maternal, portly
woman draped in flowing robes, crowned with a diadem, and
holding a torch in one hand:

> Uncle Sam symbolizes the government: demanding, negotiating,
> asking for sacrifices. The Goddess of Liberty, or Columbia, repre-
> sents the land of freedom and opportunity: America as a bountiful
> cornucopia. The national iconography seems to be sensitive to
> what America offers her people and the world as well as to what
> the goverhment requires of its citizens.[6]

The observations of two foreign visitors—the Frenchman
Alexis de Tocqueville, who visited in the 1830s, and the En-
glishman James Bryce, who visited here a half-century later—
are also instructive. Tocqueville observed that individualism
and idealism were somehow just as characteristic of the American style as conformity and materialism. He was surprised by
two facets in 1831: "The mutability of the greater part of human

actions, and the singular stability of certain principles." Bryce observed that, while Americans were shrewd and tough-minded, they were also very impressionable; while they were restless and unsettled, they were also rather associative. "Although the atoms are in constant motion, they have a strong attraction for one another." Despite their inclination to change mood and place so readily, Bryce also noted the power of habit, tenacity, and tradition among Americans, and thus concluded that "it may seem a paradox to add that the Americans are a conservative people."[7]

Hence, the inevitable question: Can a people so diverse, so polarized, and so contradictory ever manage to have it both ways—to commit themselves to a sense of community at the same time they honor the individual? Kammen, as well as numerous other students of American civilization past and present, provides a troublesome answer by observing that racial discrimination, civic corruption, and violence are just as much American traditions as equality, morality, and the rule of law. He writes:

> By constantly breaking with the past and conforming to transitory norms and fashions, the American seems to be both anti-traditionalist and highly conformist. He is also embarrassed by another moral dualism: the conflict between high ethical standards and the ethos of the market place.[8]

A more recent visitor, the philosopher Jacques Maritain, focused on the counterpoint that is the theme of this chapter when he wrote: "The feelings and instinct of community are much stronger in this country than in Europe . . . the result of which is a tension, perpetually varying in intensity, between the sense of the community and the sense of individual freedom." Maritain believed that such tension is normal and fecund in itself, but when the community feels that it is threatened a counteraction tends to follow, "in the name of moral tenets such as individual freedom and civil rights."[9] Yet, again paradoxically, the nation seems unable to hold its very existence and unity without the tension between community and individualism.

Americans fear having to give up their dream of personal success for a more collective, and thus less private, endeavor. They fear a loss of their distinctness, separateness, and individuality. The alternative, they believe, is dependence and tyranny.

This is also the theme of J. Anthony Lukas's new book, *Common Ground*, in which he examines, in his words, one of the deepest divisions in American life—between the demands of equality and the call of community:

> It is a conflict as old as the nation itself—between the notion of community expressed by John Winthrop when he set out to found a "city upon a hill" in the Massachusetts Bay Colony, and the idea of equality enshrined in the Declaration of Independence.[10]

The dichotomy is tellingly revealed through a conversation Lukas overhead in a Boston tavern "one dusky evening not so long ago":

> First Drinker: "So I told him I'm an American, you know, and I got as much rights as anybody."
> Second Drinker: "Yeh, but you're a Townie first. The Townies don't take flapdoodle off any man."

It is that sense of community, coupled with an equally strong sense of individualism, that perhaps can serve as a frame for the recognition, viability, and preservation of personal privacy. If it is true, as the evidence indicates, that Americans have always lived a life of contradiction, a style of opposites, then a solution may be found in the polarity itself. Privacy is a matter of collective resolution, not a moral matter of impersonal society, nor a legal matter of bureaucratic government. As the authors of a recent study suggest:

> What we find hard to see is that it is the extreme fragmentation of the modern world that really threatens our individuation; that what is best in our separation and individuation, our sense of dignity and autonomy as persons, requires a new integration if it is to be sustained.[11]

Unfortunately, much of the talk today about the loss of privacy really centers on economic freedom but not necessarily on economic equality. This is surely the main message of Lukas's chronicle of three American families during the turbulent decade from 1968 to 1978: their sense of alienation when the larger community became less responsive to their individual plight; their sense of frustration when public idealism shifted to private interest. Community, however defined or constructed, is the place where the individual thrives, but the community must be so integrated so as to nurture the dignity and autonomy of its members. And here is where history teaches a valuable lesson.

Community and *individual,* as descriptive terms, have been in the English language since the fourteenth century, the latter for an even much longer period of time. *Community* became established in a range of senses, as traced by the British scholar Raymond Williams: 1) the commons or common people, as distinguished from those of rank; 2) a state or organized society, in its later uses relatively small; 3) the people of a district; 4) the quality of holding something in common, as in "community of interests," "community of goods;" and 5) a sense of common identity and characteristics.

In the first three senses described by Williams, *community* indicated actual social groups, but in the others the term indicated a particular quality of relationship, as in the Latin *communitas,* meaning common possession or participation. From the seventeenth century, and especially since the nineteenth, *community* was felt to be more immediate than *society,* but both, Williams says, were an effort to distinguish the body of direct relationship from the more organized, and formal, sense of *realm* or *state.*

Beginning with general industrialization in the nineteenth century, Williams notes, *community* developed a sense of immediacy or locality in the context of larger and more complex societies:

> *Community* was the word normally chosen for experiments in an alternative kind of group-living. It is still so used and has been joined, in a more limited sense, by *commune.* The French *commune*—the smallest administrative division—and the German *Ge-*

meinde—a civil and ecclesiastical division—had interacted with each other and with *community*, and also passed into socialist thought and into sociology to express particular kinds of social relations.[12]

Increasingly, through the nineteenth century, the distinctions between *community*, as symbolic of more direct, more total, and therefore more significant relationships, and *state* or *society*, as indicative of more formal, more abstract, and more instrumental relationships, became pronounced, as the distinctions remain to this day. "Community politics," for example, is used to distinguish it both from "national politics" or even "local politics" and, according to Williams, usually involves various kinds of direct action and direct local organization, "working directly with people," but is different from "service to the community," normally associated with voluntary work. Williams continues:

> The complexity of *community* thus relates to the difficult interaction between the tendencies originally distinguished in the historical development: on the one hand the sense of direct common concern; on the other hand the materialization of various forms of common organization, which may or may not adequately express this. *Community* can be the warmly persuasive word to describe an existing set of relationships, or the warmly persuasive word to describe an alternative set of relationships. What is most important, perhaps, is that unlike all other terms of social organization (state, nation, society, etc.) it seems never to be used unfavourably, and never to be given any positive opposing or distinguishing term.[13]

Individual, the second relevant key term, originally meant indivisible, but has come to mean something quite different, stressing a distinction from, rather than a connection to, others. Williams notes that the transition was best marked by use of the phrase "in the individuall," as opposed to "in the generall." For most of the seventeenth century, the term was often used in a pejorative sense, implying a vain or eccentric departure from the common ground of human nature. Eventually the word served to classify groups of people, as individuals, families, or commonwealths, but was still used as an adjective, too, as in "our

Idea of any individual Man." *Individual* as the singular noun appeared first in logic, then in biology, as a method of classification—genera into species and species into individuals.

A crucial shift in attitudes took place in the eighteenth century, as in Adam Smith's "among the savage nations of hunters and fishers, every individual . . . is . . . employed in useful labour." Darwin recognized in 1859, in *Origin of Species*, that "no one supposes that all the individuals of the same species are cast in the same actual mould." Williams writes: "Increasingly the phrase 'an individual'—a single example of a group—was joined and overtaken by 'the individual': a fundamental order of being."[14] But, while the individual gained status in the eighteenth century, the result of a new emphasis on a man's existence beyond his place or function in a rigid hierarchical society, "the individual" as a starting point was criticized in the nineteenth century. The species was more important, according to Burke, Marx, Darwin, and Freud.

Williams maintains that many arguments about "the individual" confuse the distinct senses to which "individualism" and "individuality" point. The latter term has the longer history and developed its modern usage with the breakup of the medieval social, economic, and religious orders. "Individuality" comes out of the complex meanings in which "individual" emerged, stressing both a unique person and his indivisible membership in a group. "Individualism," coined in the nineteenth century, is based upon a theory not only of abstract individuals, but of the primacy of individual states and interests. In its modern usage, it is directly related to "private" and "privacy," as discussed in Chapter 1.

Barrington Moore, Jr., in his review of basic anthropological perspectives on privacy, writes: ". . . the conflict between the desire for independent and even 'selfish' behavior and the objective need to depend on others remains a central aspect . . . in *any* human society with *any* painful obligations."[15] Moore is in the tradition of utilitarianism when he asserts that the need for privacy amounts to a desire for socially approved protection against social obligations. Privacy thus becomes a negative right. As discussed in the previous chapter, Bentham and others argued that rights and obligations, or duties, are correlative. For them the idea of a right was regarded as superfluous, and all that was required of the individual and society was the acceptance of the idea of obligation, or duty. Rights have had a historical relationship with official law, whereas duties and obligations have been related to the quality of life in general.

Personal privacy, or independence, and social community, or interdependence, are so intertwined as to be inseparable concepts; however, distinctions between public and private spheres are considerably less visible in smaller, more primitive communities, than they are in larger, more complex societies. Moore believes that the character of a society's obligations will determine its needs and opportunities for privacy. These obligations, in turn, are derived from the nature of the social and physical environment, the state of technology, the division of labor, and the system of authority. Presumably, a democratic society, regardless of its numerical size, is likely to have fewer imposed, or even understood, obligations, and thus is freer to honor the desire for personal privacy on the part of its citizens.

"Even a strong yearning for privacy can evidently evaporate in the face of an acute awareness of one's dependence on other human beings," writes Moore. Also, the yearning can be eliminated by a need for dependence that comes from an awareness of individual helplessness and isolation. The yearning for privacy and the desire for community are of equal strength, but the need to rely upon others, even those who share the intimacy of a small group of friends or relatives, is usually of short duration. As Moore concludes: "We can expect to find alternating cycles of private dependence and public performance of demanding roles in a good many societies."[16] However defined and in whatever context, privacy in discussion is yet another way of reviving the ancient debate—the relationship of the individual to the larger community.

In classical Greece, privacy carried negative overtones and

implied the absence of full participation in the approved social order. *Idios*, the Greek word, means "one's own, pertaining to one's self," hence private or personal. *Demios*, Greek for public, means literally "having to do with the people." The English word "idiot" comes from the noun form *idiotes*, but its primary meaning is a private person, or an individual in private station as opposed to one holding public office or participating in public affairs. Another widely used term for public, *koinos*, means "common," as in shared in common as different from private property or private interest. Greek usage of the words expresses a certain bias against what is private and a bias in favor of sharing what is public. There is, however, a negative association of "public" with "common" in the sense of "vulgar and inferior."

Hannah Arendt explained how the rise of the social realm (in addition to the long-established public and private realms) has affected modern concepts of privacy:

> The emergence of society—the rise of housekeeping, its activities, problems, and organizational devices—from the shadowy interior of the household into the light of the public sphere, has not only blurred the old borderline between private and political, it has also changed almost beyond recognition the meaning of the two terms and their significance for the life of the individual and the citizen.[17]

Arendt believed, for instance, that we would not agree with the Greeks that a life spent in the privacy of "one's self" (*idios*), outside the world of the common, is "idiotic" by definition. Or, with the Romans, to whom privacy offered but a temporary refuge from the business of the *res publica*.

> We call private today a sphere of intimacy whose beginnings we may be able to trace back to late Roman, though hardly to any period of Greek antiquity, but whose peculiar manifoldness and variety were certainly unknown to any period prior to the modern age.[18]

The shift was not merely one of emphasis, however. In the ancient world the "privative trait of privacy," as Arendt de-

scribed it, was all-important, meaning quite literally a state of
being deprived of something. "A man who lived only a private
life, who like the slave was not permitted to enter the public
realm, or like the barbarian had chosen not to establish such a
realm, was not fully human."[19] Today we no longer think primarily of deprivation in our use of the term "privacy," and this
is the result somewhat of the great enrichment of the private
sphere through modern individualism.

The private life is at least as interesting and fulfilling, if not
more so, as the old public life. But what seemed to Arendt to be
even more important is that modern privacy is at least as sharply
opposed to the social realm—"unknown to the ancients who
considered its content a private matter"—as it is to the political
realm.

> The decisive historical fact is that modern privacy in its most rele-
> vant function, to shelter the intimate, was discovered as the op-
> posite not of the political sphere but of the social, to which it is
> therefore more closely and authentically related.[20]

In classical Athens the emphasis was quite firmly on public
obligation, not private rights. On the other hand, as Barrington
Moore, Jr., explains, citizenship carried definite privileges in
relation to the rest of the population. This seemingly ideal situation may be attributed to the fact that Athenian democracy,
unlike most modern adaptations, came into being without any
carefully enunciated bill of rights designed to protect the individual citizen against arbitrary abuse of the state's power. Athenian democracy triumphed without a struggle against royal ab-
solutism, which itself did not exist. Property was protected; men
were not condemned to death without a trial; a citizen was not
tried twice for the same offense; and officials were not permitted
to enter a private dwelling against the owner's will.

Although such high principles may not always have been
adhered to in practice, the Athenians erected a protective shield
against arbitrary injustices by public officials and institutions.
They had far fewer constitutional rights than most modern de-
mocracies, but the private life was nonetheless honored, al-

though more restricted than private life today because of a more
fundamental commitment to public obligations. Greek literature
is filled with examples of the ordinary citizen yearning to be let
alone to enjoy his simple private pleasures of drinking, eating,
and wenching, or to be allowed to enjoy the privacy of work.

The Greek tradition of privacy lives on, as poignantly told in
Nicholas Gage's 1983 biography of his mother, *Eleni*, in whose
mountain village the inhabitants watched over each other in the
face of constant danger from outside. Familial privacy was honored, but community survival was of greater importance. Gage
writes:

> Cut off from the rest of the world and driven together by the need
> to survive, the peasants of the Mourgana villages had no privacy.
> Every house looked into the one below. Voices traveled for miles
> through the thin air, and wherever they walked on the mountain
> paths, they felt eyes watching them from above or below or from a
> neighboring cliff. Despite the vastness of their universe under the
> sky, the villagers knew that everything they did was observed and
> overheard.[21]

The description also illustrates the paradox of privacy, especially
evident in small communities, where privacy and survival are
different sides of the same coin—interdependent facets of community life. The duality then, as now, is common to village life,
but for Greeks today it remains an integral part of their ancient
tradition.

This simple form of apolitical yearning for a private existence
occurs in many societies, as Moore's interesting study shows:

> It is not something to be treated with condescension by political
> thinkers whose answers don't work. Nor for that matter is the
> yearning a search for a purely private existence since it usually
> includes a willing participation in local forms of social life based on
> economic cooperation, kinship, and religion.[22]

Our own rush to legalize private activities, to create a right to
privacy and ever more zones of privacy, may be more a manifestation of our frustrations with officialdom, the bureaucracy
that now overshadows democracy, than it is a private person's

deliberate neglect of public issues. Perhaps, as Moore believes was the case in classical Athens, the individual would be more open to society's obligations were the society more open to his genuine participation. The Athenian experience is not without relevance through time.

"All greatness of character," wrote James Fenimore Cooper in *The American Democrat* in 1838, "is dependent on individuality," and he was dismayed that individuality did not thrive in American democracy. Individualism in Cooper's time was not especially an American trait; it was a trait of the European bourgeoisie, the middle class that had yet to appear in America. The wish to rise beyond one's original social class or stratum was relatively rare in the United States. As the contemporary historian John Lukacs points out, the pressure to conform, in possessions, consumption, habits, speech, as well as in thought allowed little room for genuine privacy, including private aspirations. "Most Americans," Lukacs writes, "wished to rise in the world: but they wished to rise within their communities, within their neighborhoods, within their groups—within them, not outside them."[23] It is indeed ironic that today's Americans, who have made personal privacy into a near fetish, a cult of the individual, have forsaken a sense of general privacy for a sense of personal need that results in a kind of isolated self-indulgence and sanctuarial self-protection.

For Americans the private life and the public life have become totally opposite realms, the former a shelter for the intimate, a repose beyond the reach of government; the latter is filled with obligations and duties too numerous to contemplate. It is in the social life, the third realm identified by Hannah Arendt, where severe ambivalence exists. Americans vigorously object to invasion by the "public sector"—Big Brother in their midst—but they seem willing to tolerate intrusion by the "private sector." That is until recently when they learned that their "private" credit card numbers are the means of an invasion as menacing as their "public" Social Security numbers. In both instances, as explored in later chapters, the perceived threat is to personal information, and the private sector is as guilty of this type of invasion as the public. The person who refuses to apply for a check-cashing card from a major supermarket chain because it requires the disclosure of personal information is rebelling against private-sector intrusion.

Bureaucratized life has become yet another complication for the twentieth-century individual in search of privacy. Social historians have noted that the overwhelming bureaucracy involves not only government and private-sector economy and industry, but the private lives of people as well—what they are doing in their homes after hours as well as what they are doing at work. Lukacs maintains that the "togetherness" of the 1950s was short-lived and rested on insubstantial foundations. "The social life, entertainments, leisure time and leisure activities of an increasing number of people were shaped by their corporate affiliations," writes Lukacs. Americans created another paradox for the individual. While they continued to seek personal privacy, symbolized by the well-adjusted family, they also grew more and more dependent upon public standards of behavior. "This publicization of private family life meant a fatal weakening of what a family should have meant," according to Lukacs. "For it is in the nature of human society that a family is a private and an exclusive element, whereas corporate institutions and the state are forces of inclusion." In neither world, the private nor the public, is the individual readily satisfied, and the compromising social life, as refuge, becomes pure hedonism. What is lost ultimately is a sense of community, where personal privacy, ironically, has always flourished.[24]

Individuality is the problem, therefore individuality is the solution. Trite but true. If we can but retain some sense of the worth of individuals at the same time we regain a sense of common welfare, the pursuit of privacy may be realized in its most fundamental way, as solitude on demand, as family intimacy, and as selective anonymity. Individuality arose from the ashes of Old World values of authority, obligation, and the com-

munal ethic. In their place, as David Flaherty learned from his
colonial study, individualism became more viable because of
America's plentifulness and independence, which suggested an
impetus to the need and search for personal privacy. Any attempt to suppress the individual, even the few deviant citizens,
was viewed automatically as an intrusion upon private activity.

Even the church in colonial America was not seen as a threat
to privacy, for the social benefits of required attendance on the
Sabbath tended to offset any permanent harm to personal privacy. Many of the pressures applied by the church to assure
conformity to moral doctrine invaded privacy from time to time,
but religious discipline probably served to benefit privacy as
well.

In Flaherty's view, church members were less likely to engage
in behavior that would jeopardize their privacy, perhaps because of the latent threat of mutual observation, but also because of the colonists' strong commitment to community survival and security. Self-sacrifice and self-realization, although
antithetical, went hand in hand in early America.

The "perilous conditions" (Flaherty's words) of late twentieth-century America, where government is seen as a major
threat to personal privacy, were far less perilous in the eighteenth century. Colonial lawmakers justified some intrusions
into private lives in the interest of public security and orderli-
ness, but the stronger influence by far was the authoritarian
world-view of Puritanism. So-called "pretend rules," as distinguished from genuine law, also encouraged a high degree of
self-discipline, perhaps out of fear of the authorities but, one
suspects, out of mutual agreement and a sense of community as
well. "Whatever the laws may have stated," writes Flaherty,
"persons had to engage in some flagrant violation to attract the
attention of the authorities in most cases. If an individual chose
to engage in antisocial behavior, this preference could mean
forfeiting his usual enjoyment of privacy. If a person lived up to
certain minimal norms, the law ignored him; for that matter the
law had great difficulty in finding out if he did not."[25]

Raymond Williams, who has written extensively on the British experience with community during the eighteenth and on into the nineteenth century, associates privacy with the government's land enclosure movement. Enclosure, which began in the twelfth century and was virtually complete by the end of the nineteenth century, was the process by which common fields, meadows, and pastures gave way to the pattern of small hedged fields and consolidated farms. Debate over the effects of enclosure still goes on, but Williams believes that the old way of life tied to community property was substantially altered, from one of general dependence to occasional private independence. Private property, as well as private legal rights, thus emerged in the face of fences and walls and in a more stratified economic system of landlord, tenant, and laborer. "Good fences make good neighbors" became the accepted standard, but the poet also observed the equally enduring notion, "Something there is that doesn't love a wall."[26] The process of privatizing public lands was slowed considerably during the second half of the nineteenth century with the enactment of legislation to preserve common land for public enjoyment.

Andrew Jackson Downing, the great American architect and landscape designer of the nineteenth century, lamented that the "Anglo-Saxon race" was disinclined to social intercourse or unrestrained public enjoyment. Downing found France and Germany abundant in public grounds, "pleasant drawing-rooms of the whole population," where people gained "health, good spirits, social enjoyment, and frank and cordial bearing towards their neighbors, that is totally unknown either in England or America." He described the public garden at Frankfurt as "one of the most delightful sights in the world," adorned neither by gates nor fences, and that "you will no more see a bed trampled upon, or a tree injured, than in your own private garden here at home!"[27]

Downing's theory was that out of this enjoyment of public grounds, a classless experience, "grows also a *social freedom*, and an easy and agreeable intercourse of all classes, that strikes an

American with surprise and delight." He went on to observe that his countrymen, in 1848, were "more and more inclined to raise up barriers of class, wealth, and fashion, which are almost as strong in our social usages, as the law of caste is in England." Downing, believing such a system unworthy of Americans, called it the meanest and most contemptible part of aristocracy: "We owe it to ourselves and our republican professions, to set about establishing a larger and more fraternal spirit in our social life."[28]

Many writers have expounded persuasively on the virtues inherent in the individual and the values associated with individualism (from John Locke to Ralph Waldo Emerson to Robert Nozick), but few have been as persuasive on the virtues and values of community as Alexis de Tocqueville and Hannah Arendt. The literature leans heavily in the direction of the individual, as opposed to the collective, which itself is a fascinating piece of lexicology because, as Raymond Williams noted, the term "community" never seems to have been used unfavorably. Individualism, on the other hand, has been frequently criticized, even by Emerson, who saw the phenomenon becoming cultist and counterproductive to the common good.

Tocqueville's idea of "individualism," rooted in egoism, was not to be compared with the ethic of "self-reliance" advocated by Thoreau and Emerson. Instead, *l'individualisme* described the diminishing power of the individual who becomes preoccupied with the self—what today's social scientists would call "privatization" or "narcissism." Individualism "disposes each citizen to isolate himself from the mass of his fellows and withdraw into the circle of family and friends; with this little society formed to his taste, he gladly leaves the great society to look after itself."[29] Egoism is worse by far, for it springs from blind instinct, whereas individualism is based on misguided judgment. "It is due more to inadequate understanding than to perversity of heart." In the long run, individualism merges in egoism and attacks and destroys public virtues.

Throughout his great and enduring analysis of the American

character, Tocqueville, too, was ambivalent about the source and effects of privacy. In one of his more perceptive observations, Tocqueville noted that democracy tends to isolate individuals, perpetuated by hatreds held over from an earlier time of inequality. Such was the kind of democratic revolution his native France had experienced. But Americans had the great advantage of attaining democracy without such extreme suffering—they were born equal instead of becoming so. It was left for Americans to devise a way to combat individualism, the utilitarian principle that Tocqueville called "the doctrine of self-interest properly understood." Then, individualism as selfishness was not openly condoned; now, individualism as personal privacy, although more acceptable, creates the same sort of ambivalence Tocqueville described more than a century and a half ago.

When the world was under the control of a few rich and powerful men, the official doctrine of morality was that one should do good without self-interest, "as God himself does." With the rise of democracy and egalitarianism, every man's thoughts centered on himself and the idea of sacrifice succumbed to private advantage. Privacy, an eighteenth-century innovation, according to Fernand Braudel, came into full flower in the nineteenth. But Tocqueville saw emerging in America another kind of virtue: "Inhabitants of the United States almost always know how to combine their own advantage with that of their fellow citizens." Tocqueville credited Americans with a strong sense of sacrifice, a willingness to relinquish their private interests to save the rest. He predicted that private interest would become the chief driving force behind all behavior, but he warned that its value to society will depend upon how each man will interpret his private interest.

> If citizens, attaining equality, were to remain ignorant and coarse, it would be difficult to foresee any limit to the stupid excesses into which their selfishness might lead them, and no one could foretell into what shameful troubles they might plunge themselves for fear of sacrificing some of their own well-being for the prosperity of their fellow men.[30]

Tocqueville feared that, instead of democratic citizens ultimately coming "to live in public," they will in the end only form very small private coteries.

Other midcentury voices were no less pessimistic about the effects of privacy on a sense of community. Benjamin Disraeli, in a speech in Yorkshire in 1844, criticized England for its divided society,

> a body of sections, a group of hostile garrisons. Of such a state of society the inevitable result is that public passions are excited for private ends, and popular improvement is lost sight of in particular aggrandizement.

In *Sybil,* published the following year, Disraeli warned that the cult of home was the enemy of community. Stephen Morley, the journalist-agitator of this popular political novel, says that individual influence is no longer a viable renovator of society:

> In the present state of civilization, and with the scientific means of happiness at our command, the notion of home should be obsolete. Home is a barbarous idea; the method of a rude age; home is isolation; therefore anti-social. What we want is Community.

Walter Gerard, Sybil's father, agrees but not completely: "I like stretching my feet on my own hearth." Notwithstanding the novelist's license to overstate his case, Disraeli's message is clear: public commitment that does not sacrifice the solitude of privacy is to be desired.[31]

Tocqueville was able to perceive real virtue in the American biformity of private interest and public welfare, although it remains to be seen how severely our twentieth-century obsession with privacy has affected such traditional community virtues as loyalty, generosity, and cooperation. As if to substantiate Tocqueville's concern, Peter Steinfels recently wrote:

> With each self constituting its own moral universe, our goals and values are ultimately no more than arbitrary preferences. The ideal of public service in pursuit of the common good is subsumed in either a nostalgic vision of small-town harmony or the tough-minded talk of self-interest.[32]

For Warren Johnson, America needs to return to traditional values in an age of scarcity, and he insists that for a sustainable economy to work it must be supported by the individual's willingness to practice self-restraint for the benefit of others.[33]

Hannah Arendt acknowledged that the rise of modern individualism had largely removed the aspect of privation from the term "privacy," but she nevertheless believed that a life excluded, either by choice or by design, from the public realm is still a deprived life. The term "public," she said, signifies two closely related but not altogether identical phenomena: first, that everything that appears in public can be seen and heard by everybody and has the widest possible publicity; second, that public means the world itself, "in so far as it is common to all of us and distinguished from our privately owned place in it."[34]

Alluding to slaves who remained shadowy types until they became free and notorious persons, Arendt wrote:

> Compared with the reality which comes from being seen and heard, even the greatest forces of intimate life—the passions of the heart, the thoughts of the mind, the delights of the senses—lead an uncertain, shadowy kind of existence unless and until they are transformed, deprivitized and deindividualized, as it were, into a shape to fit them for public appearance.

She believed no less in the private life, however, but it is through the "presence of others" that we learn the "reality of the world" as well as the "reality of ourselves," where intimacy is fully developed.[35]

Yet there are a number of things that obviously cannot stand the "implacable bright light of the constant presence of others," such as love distinct from friendship, which can be extinguished if displayed in public. And there are other "small things" that are relevant to personal privacy, but cannot be accommodated by the public realm, or the community sphere. Nor should they be regulated, the reason being that they are indeed private—not beyond community concern, but certainly beyond government intrusion and manipulation.

"Public" to Arendt also signifies the world itself, the world of "human artifact," where, as around a table where all men sit, each reposes but the table separates each of us at the same time:

The public realm, as the common world, gathers us together and yet prevents our falling over each other. What makes mass society so difficult to bear is not the number of people involved . . . but the fact that the world between them has lost its power to gather them together, to relate and to separate them.[36]

What is especially compelling about Arendt's argument on behalf of the public life, which is also the argument for a sense of community, is that its roots are in the Greek *polis* and the Roman *res publica*, both of which contained (and allowed for) the privacies of life at the same time they augured for the futility of individual life. These ancient civilizations had it both ways, worthy models for the modern age as well. What the modern world has forsaken, according to Arendt, is the ability—indeed, the desire—to accommodate diversity, on the one hand, and commonality, on the other, public commitment and private interest:

Only where things can be seen by many in a variety of aspects without changing their identity, so that those who are gathered around them know they see sameness in utter diversity, can worldly reality truly and reliably appear.[37]

The sameness of the common object, or objectives, of the community can only survive if it, the community, fails to recognize the "many aspects in which it presents itself to human plurality." Thus, the individual and the community are interdependent. When individual privacy leads to isolation, as it understandably can in modern mass society, people are deprived of seeing and hearing others, of being seen and heard by them. "They are all imprisoned in the subjectivity of their own singular experience, which does not cease to be singular if the same experience is multiplied innumerable times."[38] It follows, therefore, that to be obsessed with privacy is to live a life deprived of objective relationships, devoid of any intermediary element, and destructive of aspirations more permanent than life itself.

Community, in the sense of "direct common concern" and "common organization," as traced by Raymond Williams, turns out to be the best place for personal privacy to thrive, indeed, to survive. Our modern society, by encouraging competitiveness among otherwise egalitarian individuals, the privatizing of persons, actually creates a high degree of competitive indifference that works against common progress. When such a society promotes isolation, it fosters alienation among its people. Williams describes the alternative choice when he interprets the poet's vision:

> Wordsworth saw that when we become uncertain in a world of apparent strangers who yet, decisively, have a common effect on us, and when forces that will alter our lives are moving all around us in apparently external and unrecognisable forms, we can retreat, for security, into a deep subjectivity, or we can look around us for social pictures, social signs, social messages, to which, characteristically, we try to relate as individuals but so as to discover, in some form, community.[39]

The present yearning for privacy instead of community is rooted in an America perceived by many to be out of control and whose citizens see themselves as being manipulated by forces beyond their control. Richard D. Brown, the distinguished historian, puts it this way:

> The social order possesses so much fluidity that people live in a spiral of aspirations that is satisfied only intermittently, and never satiated. The range of possibilities is continuous, and people tend to set their sights even higher. The aspirations of modern Americans are so open-ended that they seem to thwart fulfillment.[40]

Another perceptive critic, John Lukacs, sees a connection between increasing pollution of the material world (the "environment") and the pollution of minds: of increasing carelessness, a weakening of moral convictions, civic responsibilities, and a sense of community. The former has been consequence, rather than cause, of the latter.

People today are afraid of being manipulated and worry about their personal privacy. As explained in the next chapter, notions of personality have changed dramatically, from a sense of self-mastery and individual supremacy in the nineteenth century to an overwhelming degree of self-manipulation in response to enormous social pressures at the present time. Our defense is to withdraw into self-aggrandizement, often with the help of drugs or tranquilizers. Privacy has become a form of retreat for a population that is already dangerously close to Huxley's *soma*-inspired world.

The sense of community is viewed here as both a haven for the necessary private life and as a place for creative public commitment. It is not seen as either a nostalgic return to small-town harmony or as a compromise for the current tough-minded self-interest movement. "We are all in the same fix," Richard H. Rovere said, "and we all have to strike the same balance between our need for others and our need for ourselves alone."[41] There is no turning back on advancing technology, nor slowing down to any measurable degree the growth in population, nor, finally, reducing the complexity of the world.

In the beginning, America was dedicated to the sovereignty of the individual and the establishment of a social order that now, thanks mainly to technology, functions largely on the interdependence of individuals. This is one of those paradoxes Americans have grown to endure, even relish; and today we still want it both ways. We can control privacy by controlling our mode of existence, and if we can never accomplish complete mastery, we can at least attempt an approach to it. The community approach is far less expensive than having to choose between self-interest and self-sacrifice, or between rugged individualism and cooperative utopia. An old Russian proverb expresses the community idea:

> In a field of wheat, only the stalk whose head is empty of grain stands above the rest.

7

Copyright of Personality

Elizabeth Taylor and Frank Sinatra are among a growing number of famous and not-so-famous individuals who have sought to protect their privacy through the use of a relatively new legal concept known as the "right of publicity"—the exclusive right to make money from one's popularity and prominence. Strict privacy law as discussed so far does not provide the protection they seek. The courts have consistently ruled that the First Amendment protects instead what is said about such public figures in the interest of free speech, unless, of course, it is a deliberate or reckless falsehood. As reported in the *New York Times*, "What Elizabeth Taylor wants to know—and what the law doesn't clearly tell her—is whose life is it, anyway? More specifically, can ABC go ahead and produce a 'docu-drama' based on Miss Taylor's life even over the movie star's vociferous objections?"[1]

In the case of Sinatra, his suit against an "unauthorized" biographer, Kitty Kelley, was tempered somewhat by the singer's open friendship with politicians, including some residents of the White House, who are normally beyond the protection of privacy law. His suit sought punitive damages for Miss Kelley's "misappropriation of name and likeness for commercial purposes." The suit, which was similar to Taylor's and those of other celebrities, such as Joe Namath and Clint Eastwood, was based on the right of publicity premise that nobody else can write "The Sinatra Story" without his permission. In both situations, the basic issue is actually twofold: that Taylor and Sinatra own their personalities and that the docu-drama and the biography would present them in a false light.

They contend that their life stories are commercial property and, furthermore, that they alone have the right of exploitation. Anyone else who uses the material without permission is misappropriating personal property. "Someday I will write my autobiography, and perhaps film it, but that will be my choice," Taylor is quoted as saying. "By doing this, ABC is taking away from my income." Common law right of publicity, for many years a part of evolving privacy law, asserts that an individual can control the commercial use of her or his name or image. It would be illegal, for example, to use Elizabeth Taylor's name or picture without her permission in an advertisement for a brand of cosmetics. With regard to "false light," the second issue involved, courts have held that a wrong impression of a person, even though it may not be unfavorable, is an invasion of privacy.

Right of publicity as a common law tort separate from the right of privacy has its roots in the famous Warren and Brandeis article of 1890, as does all privacy law. As we have seen, the two attorneys sought a plaintiff's right "to be let alone," and they asserted that the essential injury was to one's feelings with consequent mental anguish. They argued that, although some aspects of personal privacy may involve ownership or possession (as in traditional rights of property, contract, and trust), privacy meant to them "inviolate personality." What they sought was a way to protect the personality itself. Theirs was a remedy for the anguish caused where there had been no injury to property or breach of contract. Warren and Brandeis urged that the scope of jurisprudence be extended to protect not the reputation or the character (already covered by defamation law) but the *feelings* of individuals subjected to some form of unauthorized and unwanted intrusion. The rest of this chapter traces the history of that legal concept and suggests that it is a reasonable way to protect the commercial aspect of personal privacy.

"*Personality* was something we all once had" is the way Raymond Williams begins his etymology of the term. In its earliest English use, from the fourteenth to the nineteenth century, it meant the quality of being a person and not a thing.

> *Persona* (the Old French) had already gone through a remarkable development, from its earliest meaning of a mask used by a player, through a character in a play and a part that a man acts, to a general word for human being.

Person also early acquired the sense of an individual, and *personal* in the senses now recognized as "individual" and "private." *Personalitas* (the Latin) had two meanings: the general quality of being a person and not a thing; and the sense of personal belongings, which was adopted into English as *personality*.[2]

What matters in "personality," according to Williams, is the development from a general to a specific or unique quality. If we read, from mid-seventeenth-century prose, "for a time he loses the sense of his own personality and becomes a mere passive instrument of the deity," we could substitute the modern term "individuality." In the eighteenth century, Samuel Johnson defined *personality* as "the existence or individuality of any one," and soon the term acquired several uses for distinct personal identity. "She has but little personality," or he has an "overpowering personality," "strong personality," "dominant personality," "weak personality," and so forth. We may even speak of someone as having "no personality." In the twentieth century, a key word is "personalities," which identifies not only lively people but also important or well-known personalities. And, as the economic value of personality came into play, owners of this new-found commodity sought protection in the law. Williams explains the evolution:

> A *personality* or a "character," once an outward sign, has been decisively internalized, yet internalized as a possession, and there-

fore as something which can be either displayed or interpreted. This is, in one sense, an extreme of possessive individualism, but it is even more a record of the increasing awareness of "freestanding" and therefore "estimable" existence which, with all its difficulties, gave us "individual" self.[3]

Before that cycle was finished, however, the "modern" world had given rise to the development of the consciousness of self, a cultural phenomenon seen as early as the seventeenth century, according to the late Warren I. Susman. "Impulses that control human behavior and destiny were felt to arise more and more *within* the individual at the very time that the laws governing the world were seen as more and more impersonal." Susman surmised that as it became more difficult to feel spiritual life and activity immanent in the world outside the self, and as the rituals of the external church grew feebler, the needs of the inner self grew stronger.[4]

This decline in man's view of himself had actually begun much earlier, with Copernicus, who in the early sixteenth century had denied man his place at the center of the universe. Later Darwin was to assault man biologically, and then Freud drove in the last nail when he denied that man (in the traditional sense) had much control over himself. Yet, while science affected the subordination of man, the modern era elevated in man a sense of self-consciousness. The word "character" had become a part of the vocabulary in England and America in the nineteenth century. Susman believed that by 1800 the concept of character had come to define that particular modal type felt to be essential for the maintenance of the social order. The new concept filled two important functions: it suggested a method for both mastery and development of the self; and it provided a method for presenting the self to society. According to Susman, the culture of the nineteenth century was a culture of character, depicted in literature, the arts, popular music, and the like.

The importance of the self in terms of character development did not emerge as a concept apart from other key social concepts of the time, each related to the notion of character—citizenship, duty, democracy, work, building, golden deeds, outdoor life, conquest, honor, reputation, morals, manners, integrity, and, above all, manhood. "The stress was clearly moral and the interest was almost always in some sort of higher moral law."[5] Natural law, in other words. The most popular quotation that appeared in a number of nineteenth-century works studied by Susman was Emerson's definition of character: "Moral order through the medium of individual nature."[6] The concepts that gave special meaning to certain words, including "privacy," developed in concert with a myriad of other concepts and words that expressed cultural norms. They were, as utility theory would have it, correlative terms and, thus, connected rights.

Early in the present century, another vision of self began to emerge, a different emphasis on self-development and mastery, a new way of presenting the individual in society. Susman saw this as part of significant change in the social order, which he diagnosed as "American nervousness" over the rash of utopian writings, the appearance of systematic sociological and economic analysis in the academic community, the perceived need for "objective" and "scientific" data to solve social ills, and the development of psychological and psychiatric studies. These in turn suggested the need for a new kind of man to meet the new conditions.

From his study of advice manuals that appeared after the turn of the century, Susman learned that the vision of self-sacrifice began to yield to one of self-realization.

> In an important sense the transition began in the very bosom of the old culture. For it was what might be called the other side of Emerson—his vision of a transcendent self—that formed the heart of that New Thought or Mind Cure movement so important in the process from a culture of character to a culture of personality.[7]

That shift gave impetus—which continues today, if best-seller lists are any indication—to the concern for personal autonomy and privacy.

Personality was distinguished from character by a different

set of adjectives—fascinating, stunning, attractive, magnetic, glowing, masterful, creative, dominant, and forceful. One writer in 1915 clarified the distinction when he said that character is either good or bad; personality, famous or infamous. Susman then posited the important question: We live now constantly in a crowd; how can we distinguish ourselves from others in that crowd?[8]

Most of the self-help manuals published in the first quarter of the century presented readers, ironically, with the requirement that individuals should be "themselves" and *not* succumb to the advice or direction of others. Readers were encouraged to nurture self-mastery and self-development as well as to discover a method for presenting themselves in public society. Both recommendations differed from those proposed in the culture of character and they underscored the development of a new culture—the culture of personality.

Philip Rieff has depicted the influence of Freud and psychoanalysis on the new personality type, "psychological man," who, unlike his father, economic man, "has constructed his own careful economy of the inner life . . . and lives by the mastery of his own personality."[9] Susman had argued that the older vision of self as expressed in the concept of character was founded on an inner contradiction. That vision, he said, asserted that the highest development of self ended in a version of self-control or self-mastery, "which often meant fulfillment through sacrifice in the name of a higher law, ideals of duty, honor, integrity. One came to selfhood through obedience to law and ideals." But the emerging culture of personality also had its paradox. In Susman's words,

> It stressed self-fulfillment, self-expression, self-gratification so persistently that almost all writers as an afterthought gave a warning against intolerable selfishness, extreme self-confidence, excessive assertions of personal superiority.[10]

Freudian psychology had a profound impact on self-perception, but the culture of personality was also affected by the developing consumer society. The older personal and social

needs were relevant to the vision of character; the changed social order, from a producer to a consumer society, from scarcity
to abundance, was better suited to the culture of personality.
And, what is almost too perfect an irony, the social role demanded of those in the new culture was that of performer.
Every American was to become a performing self, as Susman
described the change.

The new interest in personality—emphasizing both the
unique qualities of the individual and the performing self that is
attractive to others—was not limited to self-help book authors.
Ezra Pound, among others in the high culture, pleaded for "the
rights of personality" and argued for "the survival of personality" in the modern world. As if in anticipation of the concerns
of Elizabeth Taylor and Frank Sinatra (or those of his literary
colleagues), Pound had insisted to a friend that mass society
created a world in which man was continually being used by
others.[11] Or, as Richard Schickel explained in his biography of
Douglas Fairbanks, Sr.:

> Indeed, it is now essential that the politician, the man of ideas,
> and the non-performing artist become performers so that they
> may become celebrities so that in turn they may exert genuine
> influence on the general public.

Schickel noted that Fairbanks himself was dedicated not to his
art but to himself. A famous actress said of him in 1907 that he
could be famous in films. "He's not good looking. But he has
worlds of personality."[12]

The transition from individualism to the self as character and
then to the culture of personality is remarkably coincidental
with increasing concern over personal privacy. This brief historical trip also provides some appreciation of the search for legal
protection of one's self, in this case the assurance that one's
personality will not be appropriated without consent and,
eventually, compensation. It is an example, as is apparent in the
following cases, of the law attempting to respond to the needs of

human behavior. But the trend toward legalizing the personality—probably not what Ezra Pound had in mind—is also part of yet another significant change in American culture.

Traditional American values, as well as the individuals who hold them, have been depersonalized by an impersonal society. Privacy has gone beyond the needs of prior generations; it has become the legal right of the individual, and the rich and famous believe they, too, are at the mercy of large-scale bureaucracies, public and private, which exploit their personalities. This transition, which Warren Susman and other social historians have traced so suggestively, has been aggravated even more by the increasing importance placed on monetary success of the "new" personality as against artistic or scientific acclaim of the "old" individual. Those persons who have acquired a real property value in their names stand to lose the most. Or else Raymond Williams's tongue-in-cheek etymology is serious business: "Personality was something we all once had." Personality has become a personal belonging worth money—lots of money in some cases.

Right of publicity as legal protection for personality as property emerged with *Haelan Laboratories v. Topps Chewing Gum* (1953), when Federal Circuit Judge Jerome N. Frank said that the plaintiff may be as offended by commercial exploitation as by humiliation or embarrassment in appropriation suits. Judge Frank identified "right of publicity" and argued that individuals have a right to control the commercial use of their name or likeness.

> For it is common knowledge [he wrote] that many prominent persons (especially actors and ballplayers), far from having their feelings bruised through public exposure of their likenesses, would feel sorely deprived if they no longer received money for authorizing their countenances, displayed in newspapers, magazines, buses, trains and subways. This right of publicity would usually yield them no money unless it could be made the subject of an exclusive grant which barred any other advertisers from using their pictures.[13]

Judge Frank's opinion for the majority stands as the real beginning of the legal right of personality. Its impact has been substantial.

In the late 1970s and early 1980s, the right of publicity/personality continued to move away from the traditional legal right of privacy. What appears to distinguish the two legal concepts is this: privacy focuses on how intentional distortion, called false light and fictionalization, affects the individual, particularly her or his feelings; publicity/personality focuses on what the individual may own to the exclusion of others, an economic right of personal property, in other words. Erik Lazar notes that in publicity, performance, and copyright, "it is the personality that creates the asset, and consequently the personality that is being exploited."[14] Hence, "publicity figure" has come to mean, or describe, a person who, by virtue of public exposure, generates an interest in her or his life and likeness, and includes celebrities, public figures, entertainers, and athletes. But, because they voluntarily live in the light of publicity, the law as it has evolved in this area of privacy is troublesome.

Although the heirs of actor Bela Lugosi were unsuccessful in their effort to stop a motion picture studio from exploiting the Count Dracula characterization created by Lugosi when he was under contract to the studio, the case generated four different views, or models, as to the ownership of the commercial rights to, in this instance, the actor's famed portrayal. They are: 1) property right; 2) privacy right; 3) employer-owned product of employment; and 4) copyright. Kevin Marks's assessment of the models is an important contribution to the concept of a right to publicity/personality.

The publicity figure's interest is the right to profit from his image and it is also the right to manage his image as he sees fit. As Marks posits the issue: "Unless the public figure has marketing control by license or assignment over the commercial use of his name or likeness, the asset's value is reduced because it is subject to waste by self-interested entrepreneurs."[15] The value of the property model is that it is assignable and descendable, that is, the person is able to assign his name or likeness to others for maximum exploitation, and his heirs may acquire ownership of the name or likeness.

The model's weakness is its failure to recognize the countervailing social interests in free enterprise and free expression. In a suit brought by Elvis Presley's heirs against a company selling replicas of a statue of the singer, the court reasoned that the memory, name, and pictures of famous individuals should be regarded as a common asset to be shared, "an economic opportunity available in the free market system." In other words, any recognition of a right of publicity/personality should take into account the First Amendment's strong guarantees.

As for the privacy model, it is not applicable to cases involving celebrities and public figures who voluntarily expose themselves and their exploits to public view and scrutiny. To them, public exposure is desirable, even necessary, and no legal invasion could be protected by the right of privacy. Confusion over the various court decisions may rest with whether privacy or publicity law is applied. If privacy, there may be conflict with the First Amendment's mandate for openness; if publicity, or property, law is applied, no such conflict may occur.

In considering the work-product model, one of the judges said that no proprietary rights inured to Lugosi (and hence his heirs) because it was the image of Dracula, not Lugosi, that was marketed and that an employee's creation in the course of his employment normally belongs to the employer. It is much like the distinction between authorship and performance rights under copyright law. For example, Arthur Miller, the creator of Willie Loman, owns all the rights to that character, but not so the scores of actors who have performed the role in *Death of a Salesman,* not even Lee J. Cobb, who "created" the stage personality. However, where an entertainer develops an original character, as Stan Laurel and Oliver Hardy did, the right of publicity/personality survives the death of the creator. Lugosi's fictional character had been created by Bram Stoker, the British novelist.

Finally, copyright law, a variant of property law, seeks to

secure a fair reward for creative efforts in order to achieve two related goals, that of encouragement of artistic endeavor, and assurance that the return to the creator is a fair one. There is also a third understood goal—society's interest in having access to the works of its creative citizens. The California chief justice, in dissent in the Lugosi matter, acknowledged both the artistic incentive and entitlement to the fruits of one's labors as compelling justification for recognizing the right of publicity/personality.

In *Zacchini v. Scripps-Howard Broadcasting Co.* (1976), the U.S. Supreme Court applied the copyright model, but, whereas it endorsed the first two goals, it ignored the important third in its majority opinion.[16] Despite some of its bewildering aspects for the news media, *Zacchini* is the most significant right of publicity/personality case to date. It is so because of the judicial disagreements it engendered and, also, because it includes valuable statements on privacy, free press, publicity, and copyright. The case has all of the necessary ingredients for the speculator.

Hugo Zacchini, a human cannonball, was performing at a country fair in northeastern Ohio, when, over his objections, a reporter from a Cleveland television station filmed his act. A general admission fee was charged to enter the fairgrounds, but there was no additional cost for Zacchini's performance. Reporters were admitted to the grounds without charge, and the reporter in question visited the fair on two occasions. On the first, Zacchini requested that his act not be filmed; but on his return visit the reporter filmed the performance anyway. A fifteen-second film clip of Zacchini's stunt was broadcast by WEWS during its 11 p.m. news program. The clip was accompanied by favorable commentary that even urged people to see the "thriller . . . *in person.*"

Zacchini sued for invasion of privacy, alleging that the broadcast was an unlawful appropriation of his *professional* privacy. The trial court granted Scripps-Howard's motion for summary judgment, but the Ohio Court of Appeals reversed the decision, holding that the "total appropriation" of the performer's act was an invasion of the property right "which will give rise to a cause of action . . . based either on conversion [of property] or the invasion of the performer's common law copyright." Two of the judges relied on the doctrines of conversion and copyright, the third on Zacchini's right of publicity/personality. But they all agreed that the First Amendment did not protect the broadcast.

The Ohio Supreme Court reinstated the trial-court determination, saying that, even though Zacchini enjoyed a performer's right to the publicity value of his performance, his act was nevertheless a matter of legitimate public interest, which the station was privileged to report. Rejecting conversion and copyright and citing *Time Inc. v. Hill* (1967), the court said that "freedom of the press inevitably imposes certain limits upon an individual's right of privacy."[17] The court also noted, from *New York Times Co. v. Sullivan* (1964), that the privilege could be lost only if the intent of the press "was not to report the performance, but rather to appropriate the performance for some other private use, or if the actual intent was to injure the performer."[18]

In its six-to-one decision, the court said:

> Certainly it has never been held that one's countenance or image is 'converted' [the wrongful claim over property in exclusion of the right of the owner] by being photographed.

The court also said that Zacchini's performance was "safely outside" the bounds of copyright designed to foster and protect literary and artistic expression. In reviewing similar cases, the court reaffirmed that

> the press has a privilege to report matters of legitimate public interest even though such reports might intrude on matters otherwise private. The same privilege exists in cases where appropriation of a right of publicity is claimed, and the privilege may properly be said to be lost where the actual intent of the publication is not to give publicity to matters of legitimate public concern.

Justice Celebrezze, in dissent, questioned the court's reliance on the "public interest" standard as formulated in *Time Inc.* and

New York Times in light of the subsequent "public figure" standard of *Gertz v. Robert Welch Inc.* (1974). In the last-named decision, the U.S. Supreme Court held that the extent of the constitutional privilege accorded libel defendants depends upon whether the plaintiff is a public figure, not upon the public's interest in the subject matter. Celebrezze said he believed that the *Gertz* standard may apply to false-light privacy, despite the fact that Zacchini had alleged appropriation of his economic livelihood. Celebrezze agreed that Zacchini was protected by a right of publicity/personality, but raised the question of whether he was a public figure, as defined in *Gertz*. If not, then the television station's privilege would not have been the decisive factor.[19]

The U.S. Supreme Court, in yet another puzzling reversal in *Zacchini*, said that the First and Fourteenth Amendments did not immunize the broadcasting company from liability for televising the performer's entire act. It said that *Time Inc.*, a false-light case, was entirely different from the right of publicity. Justice Byron White, for the majority, wrote:

> In "false light" cases, the only way to protect the interests involved is to attempt to minimize publication of the damaging matter, while in "right of publicity" cases the only question is who gets to do the publishing. An entertainer . . . usually has no objection to the widespread publication of his act so long as he gets the commercial benefit of such publication.[20]

White compared right of publicity to common law copyright.

> The Constitution no more prevents a state from requiring respondent to compensate petitioner for broadcasting his act on television that it would privilege respondent to film and broadcast a copyrighted dramatic work without liability to the copyright owner.

Broadcasting the entire act, he said, "poses a substantial threat to the economic value of that performance." The justice said that, if the public can see the act free on television, it will be less willing to pay to see it at the fair. But he acknowledged in a footnote that television exposure could have increased the value of the performance by stimulating public interest in seeing the act live. One suspects that had the station televised something short of the entire act, the court would have required Zacchini to show damages, as *Gertz* requires in both private-person and public-person defamation action. The Zacchini decision goes to the heart of the publicity/personality doctrine as strictly an economic matter of privacy.

> The broadcast of petitioner's entire performance, unlike the unauthorized use of another's name for purposes of trade or the incidental use of a name or picture by the press, goes to the heart of petitioner's ability to earn a living as an entertainer.[21]

Three members of the court, in dissent, tried to take into account the possible impact of the "entire performance" rule on freedom of expression. Justice Powell alluded to the rule's chilling effect:

> Hereafter whenever a television news editor is unsure whether certain footage received from a camera crew might be held to portray an "entire act" he may decline coverage—even of clearly newsworthy events—or confine the broadcast to watered-down verbal reporting, perhaps with an occasional still picture. The public is then the loser. This is hardly the kind of news reportage that the First Amendment is meant to foster. Rather than begin with a quantitative analysis of the performer's behavior—is this or is this not his entire act?—we should direct initial attention to the actions of the news media—what use did the station make of the film footage?[22]

Justice Stevens, in a separate dissent, wanted to remand the case because he doubted whether a federal constitutional right was in question. He thought the basis of the Ohio Supreme Court's decision to be "sufficiently doubtful," that it may have extended too far the reach of the common law.[23]

Zacchini is important for a number of reasons. First, the U.S. Supreme Court upheld both the value of protecting personal privacy and the specific right of publicity/personality. The Court reaffirmed its interest in the whole privacy question. Second, the case helped to establish the difference between false-light privacy and economic appropriation. Third, the Court also determined that the "public interest" standard, as applied in previous cases, is not applicable to privacy areas other than false-light. Finally, the First Amendment is not an absolute protection for even factual news reports if they infringe on a performer's right of publicity/personality. But the decision also left a number of questions unanswered.

Justice Powell's notion that an intention to exploit someone else's performance solely for commercial gain would be enough to support a finding of liability is remarkably similar, as Bennett D. Zurofsky observes, to the majority's point that the right of publicity protects a performer from the theft of her or his good will for the unjust enrichment of another.[24] Yet the majority failed to deal with "unjust enrichment," nor did the Court rebut the dissenters' conclusion that there had been none. The mere appearance on a late-evening newscast of the performance, albeit the entire fifteen-second act, hardly seems to indicate "unjust enrichment" beyond that of public interest in an unusual, and unique, daredevil stunt.

The Supreme Court has ruled in libel cases that public figures who voluntarily inject themselves into an event or issue where public interest is clearly an element must show that the offending party had acted with "actual malice." In a suit brought by a celebrated college football coach against the *Saturday Evening Post*, the Court said he had to show severe recklessness. Because the libel directly affected Wally Butts's ability to make a living, were he to litigate his case today, his attorney might suggest an action based on right of publicity/personality.[25] Justice Powell's coinage of "intention to exploit" is similar to "actual malice" in libel and may become a crucial factor in determining invasion of privacy fault in the future. One assumes that the television station's knowledge of Zacchini's objections created the requisite fault, yet "intention to exploit" is surely not the same as the responsibility to report events of normal public interest. Powell's position, although not as liberal as the news media might like, is a tolerable approach for protecting two very important democratic ideals—the individual personality and the public interest.

Perhaps what is most troublesome about *Zacchini* is Justice White's belief that right of publicity/personality is analogous to copyright, as it is also for patent, unjust enrichment, and unfair competition laws, where inevitably the concept of fair use comes into play. White insisted that the considerations underlying the right of publicity are identical to those underlying copyright! This would mean also that the right is descendable. But, if the courts remember society's utility interest in having access to the works of its creative citizens, most copyright provisions might serve the common good as well.

Any further analysis of the right of publicity/personality must take into account some of the confusion and ambivalence surrounding this recent legal concept. Warren and Brandeis, whose pioneer treatise reads today more like a historic anomaly than a viable guide to privacy protection, set out to protect an individual's right to selective anonymity—the principle that each of us should be able to control, with few exceptions, the circles within which details of our lives are disseminated to others. Neither the courts nor the mass media have been able since then to apply the concept with much success. It is simply too broad to set realistic legal limits on the dissemination of private information. As a prominent American sociologist recently observed in a somewhat different context, "For legal purposes, stable and widely applicable definitions might be desired . . . but sociologically and psychologically that is not possible."[26] In the end, it may be impossible to create, or find in existing law, "objective" terminology that applies to such a highly "subjective" problem as privacy.

Like it or not, liberal theorists, including those who espouse a utility notion of human rights, may have to adapt some economic thinking to their social planning. Here, the recent scholarship of Judge Richard Posner, as it pertains to the right of privacy and especially the publicity/personality aspect of that legal concept, is important. His economic theory of privacy, fetching for its harsh realism and seductive rationale, anticipates the next chapter on the media, for Posner is concerned with the positive role the press plays in "privacy" and "prying," the two intermediate commodities of his entire scheme.[27]

Privacy and prying are valued as intermediate goods, rather than ultimate values, because so much of either privacy or prying is a matter of "taste," like the interest we have in turnips or beer, and taste is unanalyzable as a consumption product for economic charting. In Posner's words, "People are assumed not to desire or value privacy or prying in themselves, but to use these goods as inputs into the production of income or some other broad measure of utility or welfare." Behind the copyright of personality is money not privacy. Very few people want to be let alone. What they want, Posner believes, is to manipulate the world around them by selective disclosure of facts about themselves. They simply want to be in control of information about them. "Why should others be asked to take their self-serving claims at face value and be prevented from obtaining the information necessary to verify or disprove these claims?"

Posner is not talking about personal privacy as the need for solitude and seclusion away from the demands of daily life. He is talking about privacy as personal information, not all of which is discreditable to the individual to whom it pertains, nor does it have much to do with autonomous behavior. Posner asserts boldly—and rightly—that much of the demand for privacy, however, concerns discreditable information, "moral conduct at variance with a person's professed moral standards." Prying, in economic terms, thus has as much value as privacy, for, as Posner writes, "Even a pure altruist needs to know the (approximate) wealth of any prospective beneficiary of his altruism in order to be able to gauge the value of a transfer [of information] to him."

To the accusation that Posner's economic analysis is too reductionist, in that it fails to distinguish between privacy as an end and privacy as a means, the judge responds by joining the two goals in legal, as well as economic, terms:

> With regard to ends there is a prima facie case for assigning the property right in a secret that is a byproduct of socially productive activity to the individual if its compelled disclosure would impair the incentives to engage in that activity; but there is a prima facie case for assigning the property right away from the individual where secrecy would reduce the social product by misleading the people with whom he deals.[28]

However, just because most facts about people belong in the public domain does not imply, under Posner's analysis, that the law should generally permit intrusion on private communications, "given the effects of such intrusions on the costs of legitimate communications."

Posner would protect the following kinds of privacy on the basis of economic efficiency: 1) trade and business secrets by which the rightful owners of such information exploit their superior knowledge or skills; and 2) lives threatened by eavesdropping and other forms of intrusive surveillance, including, so far as possible, surveillance of illegal activities. He advises no protection for facts about people: "Even my income would not be facts over which I had property rights although I might be able to prevent their discovery by methods unduly intrusive."

Among the several theorists who are critical of Posner's "seriously misleading" theory and who relate privacy to individuality, Edward J. Bloustein finds unnerving—and rightly so—the overemphasis on only one of the "instrumental values" privacy serves. Posner deemphasizes the use of privacy "as a condition of creativity of thought and artistic creation and its importance as a condition of forthright and effective social and intellectual communication."[29] Posner's vision may be limited, but it forces us to deal with the economic realities of privacy. More magnanimous theories, as deserving as they are, often fail precisely because they associate personal privacy with a myriad of other indices of the human condition. For example, as Posner points out in rebuttal, history does *not* teach that privacy is a precondition to creativity or individuality: "These qualities have flourished in societies, including ancient Greece, Renaissance

Italy, and Elizabethan England, that had much less privacy than we in the United States have today."[30] Furthermore, if George Steiner's use of the haunting paradox is relevant to the discussion, historial evidence, he writes, goes a long way to suggest that great art flourishes under repressive conditions. Social-political repression is not the *cause* of creativity, but neither is privacy the begetter of dignity. "The informing context of personal creation is always social and collective," Steiner writes.[31]

The sheer size and pervasiveness of the media have shattered the line between reality and myth, fact and fancy, substance and form, and, to borrow from the late Marshall McLuhan, the difference between message and media. Barbara Goldsmith has commented:

> Today we are faced with a vast confusing jumble of celebrities: the talented and untalented, heroes and villains, people of accomplishment and those who have accomplished nothing at all, the criteria for their celebrity being that their images encapsulate some form of the American Dream, that they give enough of an appearance of leadership, heroism, wealth, success, danger, glamour and excitement to feed our fantasies.[32]

What, then, is privacy law, designed to protect the personality, to do? Whom is it to protect?

Alexander Meiklejohn, who once identified "protected speech" as that which helps citizens in a democracy to govern themselves, considered the ownership of property a constitutional right limited in terms of public, or governmental, restrictions. Government may take whatever part of a man's income it deems necessary for the promotion of the general welfare. Meiklejohn said that we confuse freedom of belief and freedom of property as having the same "freedom" meaning, and, thus, we are in constant danger of giving to a man's possessions the same dignity, the same status, as we give to the man himself. This careful distinction has bearing on personality as it is viewed as a property right, that society's right to information is as strong as is the individual's right to anonymity.[33] Zechariah Chaffee noted that a major source of ambivalence about privacy versus exposure is the pleasure that many people derive from receiving publicity in the media:

> Times have changed since Brandeis wrote in 1890. Seeing how society dames and damsels sell their faces for cash in connection with cosmetics, cameras, and cars, one suspects that the right to publicity is more highly valued than any right to privacy.[34]

It may be possible in theory to separate substance from form, or, in the original cases of Elizabeth Taylor and Frank Sinatra, fact from fiction, even the private person from the public one. It is silly to assume that such celebrities, whose prominence is directly the result of vast public exposure, have the exclusive rights to their life stories. But it is also silly to say that they have absolutely no claim to personal privacy. The important point is that their status as *public* personalities allows for little real means for determining the legal boundaries of exploitation. The terms "celebrity" and "personality" are now interchangeable in our language, as Goldsmith observes, making the Taylor and Sinatra claims of ownership tenuous indeed. Roscoe Pound, once dean of the Harvard Law School, wrote many years ago, adapting William James's proposition to law, that all demands the individual may make are to be met so far as they are not outweighed by other demands of 1) other individuals, 2) the organized public, and 3) society.[35] But because it is impossible to satisfy all demands equally, there is still the need for criteria for determing which demands—those of the individual or of society—are to be met. Ronald Dworkin's principle-policy, discussed in a previous chapter, is an estimable device, but so are the early right of publicity cases, which provide some guidelines.

Roberson and *Pavesich*, treated in detail in Chapter 2, identified "commercial purposes" as a necessary element for damages. *Pavesich* said that public persons may retain some protection against what the court called "blatant exploitation," and all sought to balance the concern of the personality trying to protect her or his publicity right against the public right to know. In *Haelan Laboratories*, the court said that the right of publicity had

little to do with privacy because bruised feelings are risks prominent persons take and the price they pay for public exposure
over time. Rather than linking television docu-dramas and unauthorized biographies to commercial exploitation, the offended
personality *and* the public's right to know might be better served
if the principle of knowing or reckless falsehood were applied.
In which case, libel law, rather than exploitation or false-light,
would determine damages. Furthermore, because most plaintiffs in such situations are public figures, the malice standard
would come into play, as applied by the Ohio Supreme Court in
Zacchini.

What is also problematic about the right of pub-
licity/personality, as with the privacy tort in general, is the element of prior restraint. With defamation, an offense is dealt with
after the fact and poses little direct threat to freedom of expression. The same is true with copyright. The offending party is
protected by the First Amendment but later may be held account-
able if one's character has been defamed or copyright has been
infringed. Celebrated personalities often try to head off a possible
offense *before* any has taken place. Howard Hughes lost that
argument in 1966 when he tried unsuccessfully to stop publication of a biography about him. The court said that public interest
should be the primary concern in any fair use determination
under copyright law:

> Whether an author or publisher reaps economic benefits from the
> sale of biographical work, or whether its publication is motivated
> in part by a desire for economic gain, or whether it is designed for
> the popular market . . . has no bearing on whether a public bene-
> fit may be derived from such a work.[36]

Copyright law and the new publicity/personality tort share a
common goal: the protection and encouragement of intellectual,
literary, and artistic creativity. They are "one pea in two pods,"
as one writer put it. Both legal doctrines protect the unauthorized use of original works and allow for financial compensation.
Although they have the same end, there are significant differences between them. Right of publicity/personality is meant

to protect the creator's fame, her or his persona; copyright protects more tangible creative expression. And, as Marc J. Apfelbaum points out, under copyright law, a work is controlled by its creator, whereas under the other, control is often placed in the hands of the subject depicted in the creative work.[37]

Copyright cuts both ways: it allows for compensation to the creator and is a means for the stimulation of "artistic creativity for the general public good." The courts, in fact, have ruled that the creator's reward is secondary to the goal of encouraging literary and artistic works. Copyright law seeks to maximize societal enrichment by protecting only the author's expression of ideas and not the ideas themselves. "Through the idea/expression dichotomy," Apfelbaum writes, "copyright law provides monopolies in expression while simultaneously encouraging competition in ideas."[38]

When Congress passed the Copyright Revision Act of 1976, it took into consideration both the interests of creators and the broader interests of society. Under the revised law, the principle of fair use was statutorily enacted for the first time in U.S. history, emphasizing the government's commitment to preserving society's interest in the creative endeavors of its citizens. Fair use puts limits on the exclusive rights of copyright holders and tries to balance the interests of both parties. Fair use is determined when a work is used "for purposes such as criticism, comment, news reporting, teaching . . . scholarship or research." The common theme is that each of these fair uses is a *productive* use resulting in some benefit to the public, contrasted with an *ordinary* use, which may be deemed an infringement. Perhaps the same distinction can be applied to the right of publicity/personality when an individual's persona is in conflict with society's right to benefit.

Even celebrities have legal protection under libel law, as we have seen, if the material reported is deliberately inaccurate or untrue, and under general privacy law, if they are depicted in a false light. They also enjoy the protection of copyright law, especially when their performances or works are fixed in a tangible form. The rub comes in publicity/personality law because it covers personal attributes such as name or likeness. But that may be too broad a legal concept for the public good to be served as well. *Zacchini*, which is upsetting as First Amendment law, illustrates nonetheless why copyright is the more appropriate protection for the personality as a "work of authorship."

It may matter little in the short run whether Elizabeth Taylor, Frank Sinatra, and other celebrated personalities can prevail in the courts. If *Zacchini* is any indication of future rulings, the courts are likely to assess each case on its own merits. But what does matter in the long run is the problem such suits create for the public flow of information as implied by the First Amendment. False-light privacy and the newer publicity/personality right are attractive to celebrities because they can restrain production or publication. The prominent constitutional attorney Floyd Abrams explains the problem:

> Prior restraint is a very heavy burden on freedom of expression. Broadcasters have to build in the cost of litigation, so as celebrities become more litigious, and file more suits to try to prevent dramas or movies about them from being made, fewer people worth writing about will be written about, and the public will lose a significant means of being informed.[39]

Zacchini is a good case for the juridical verbiage it gives us to ponder. But a better precedent for establishing the doctrine that nobody's "life" is above public interest and scrutiny is that which involved the late Howard Hughes. Random House had planned to publish a biography of the famous recluse in which a series of articles about Hughes in *Look* magazine were quoted extensively. When Hughes learned of the project, he formed a corporation that purchased the rights to the magazine series, and then sued Random House for using the material without permission. The court, in ruling for the publisher, noted that the purpose of copyright law was not to stop dissemination of information about privacy-seeking public persons. The court said that it would be contrary to the public interest to allow an individual to prevent anything being written about him. It said that the law was not meant to interfere with the public's right to be informed on matters of general interest.[40]

Public benefit is almost always at risk in most matters affecting personal privacy, as is the reverse situation. But, although the courts have been reluctant to address the privacy issue boldly, various legislatures have tended to jump in with both feet and have obscured the public's right to benefit. Legislation directed at the privatization of information is explored in subsequent chapters, but it is enough here to suggest that an open democratic society cannot tolerate a high degree of privacy. Neither can it afford to become so egalitarian as to be repressive and exploitative. But the private and the public *are* connected. "Private eros and public license, the body and the body politic, interact at every level," writes George Steiner, the distinguished literary scholar.[41]

Character, once an outward sign of individualism, has become completely internalized as personality, confiscated, as it were, as a possession that can be exploited—what Raymond Williams called the "extreme of possessive individualism." The copyright of personality is a tolerable way to protect the commercial aspect of personal privacy, for there is no way of denying monetary success in a consumer society. But the law must find a way to protect public access as well.

8

Beware the Watchdog

"We consider this case against the background of a profound national commitment to the principle that debate on public issues should be uninhibited, robust, and wide-open." This is the crux of the U.S. Supreme Court's ruling in the landmark libel case, *New York Times v. Sullivan* (1964), and it is also one of the clearest interpretations of the First Amendment.

Americans accept as fact that their founding fathers meant for them to govern themselves, and, accordingly, the Bill of Rights includes the mandate to protect, in Professor Alexander Meiklejohn's words, "the freedom of those activities of thought and communication by which we govern." Judge J. Skelly Wright, in a similar but more emphatic vein, believes, "if our democratic form of government is to continue to operate successfully, there must be free, totally unfettered flow of information."[1] Yet, despite the accuracy of the statements and the high-minded quality of the rhetoric, there lingers the feeling that such is the stuff of dreams, that there are no absolute freedoms, that there is at least a delicate balance between private freedom and public welfare.

What is perhaps most crushing to the court's call for vibrant discussion and the professor's rationale for self-determination is the powerful structure and overwhelming pervasiveness of

what today are called the mass media. Although American journalists have always been at the head of the class in preserving
most democratic ideals, they have also been among the most
flagrant invaders of personal privacy. And beyond the frequent
intrusiveness of the traditional media, modern technology now
makes it possible for man to expand his post-Orwellian night-
mares to the world George Steiner describes poetically: "The
passing spy satellite picks out and registers the hermit's shadow."[2] The media are rarely so menacing of solitude and seclusion, but neither are they free of culpability in the national effort
to protect individual privacy. Three recent incidents illustrate
the basic conflict—between freedom of expression and the right
to privacy.

In overturning a ten-million-dollar verdict in 1984 against a
magazine distributor who was accused of invading the privacy
of a popular novelist, the Second Circuit Court of Appeals noted
the chilling effect of such decisions on the First Amendment.
"But a verdict of this size does more than chill an individual
defendant's rights," said Judge Richard J. Cardamone, "it deep-
freezes that particular media defendant permanently."

The suit involved the publication in *Adelina* magazine of nude
photographs of someone who was misidentified as Jackie Collins, the novelist, although the pictures were actually of an
anonymous actress appearing in a movie version of one of Miss
Collins's books. Judge Cardamone declared that "freedom of
the press and a free society either prosper together or perish
together. Yet, because of its enormous power, the contemporary
press is under heavy attack because of a widely held perception
that it uses its special First Amendment status as a license to
invade individual privacy." The judge said that "even vulgar
publications" were entitled to the protection of the Constitution.[3]

The second incident, involving a high-ranking government
official, illustrates not only the basic conflict between personal
privacy and press freedom, but it also is an example of how
standards for the private behavior of public officials are con-

nected to the mores of the society. A *New York Times* reporter poses the dilemma: "One of the most vexing, recurring problems for official Washington, and for the national press, is how to deal with reports of private misdeeds or failings of public officials."[4] The historian J. H. Plumb tells us that once upon a time the unquestioned assumption was that private matters had nothing whatsoever to do with whether a man was an excellent public administrator or a poor one. "In a world of mass media, of readily accessible copying machines . . . as privacy vanishes, the public and the private man have to become one."[5]

The press had given a lot of attention to the disclosure that John M. Fedders, chief of the enforcement division of the Securities and Exchange Commission (SEC), had periodically beaten his wife. Publicity over the couple's divorce proceedings played a major part in Fedders's decision to resign his post the day after the news reports appeared. In his letter of resignation, which was released to the press by the SEC chairman, Fedders said that the newspapers had exaggerated allegations in the divorce trial and had "unfairly described occasional highly regrettable episodes" in his marriage. He called the infrequent violent occasions during more than eighteen years of marriage "isolated events" that did not justify the "extreme characterizations made in the press." Although Fedders defended his ability to function in his job, "the glare of publicity on my private life threatens to undermine the effectiveness of the division of enforcement and of the Commission."[6]

Fedders's marital problems came to light when the *Wall Street Journal* published a long account of the divorce proceedings. The article also related that Fedders had borrowed heavily since joining the SEC and reviewed questions about his involvement in a corporate bribery case. In a follow-up report that appeared in the *New York Times* for February 27, 1985, the publicity was said to have caused embarrassment to the Reagan administration because the President condemned family violence. The *Times's* analysis of "life in the spotlight" noted that most of the time the heavy drinking of elected officials, the homosexuality of certain

members of Congress, "the casual drug use or philandering or
marital discord or financial wheeling and dealing of other officials, are discreetly ignored by their colleagues, their superiors
and the press."

Spouse abuse has moved to the top of the list of personal
offenses, a place once held by divorce, alcoholism, drug use,
homosexuality, and adultery. Society's mores have risen to a
more tolerant level on these other private matters. As the historian Plumb notes the change, "More grievous, and more dangerous to the public good, is double talk, men preaching one
sort of high-minded morality, yet practicing another. Men and
governments create a public image out of their ideas."[7] Were it
not for the admitted wife abuse in his divorce, the Fedders incident may never have drawn public attention in the press. As
John Herbers of the *New York Times* opined:

> That White House officials should agree so quickly to the depar-
> ture once Mr. Fedders's private life became public confirmed that
> wife abuse is now more unacceptable than other varieties of per-
> sonal behavior which brought resignations in the past.[8]

"This is the toughest call I've had to make since I've been in
the job," said Norman Pearlstine, managing editor of the *Wall
Street Journal*. "As a general rule I don't think we should be
writing about the private lives of public officials when there is
no indication that the behavior in private is affecting their public
performance." Pearlstine said that several factors had "prompt-
ed us to go against our general rule in the Fedders case." The
newspaper's decision was influenced by "facts"—that Fedders
was one of the most important law enforcement officials in the
country; that he admitted in public the charges of wife-beating;
that the White House was aware of the issue of family violence
and appeared to be concerned about it; and that Fedders had
said he would resign his SEC post if that would get his wife to
take him back. "When you put all that together with the questions raised about his indebtedness and some of the issues
raised in the Southland [bribery] case, it seemed that we had to
run the story."[9]

In an article in 1983, Fred F. Fielding, who was then the White House counsel, urged applicants for important government jobs not to hold back any embarrassing personal facts that could be used against them later. He said it is better to deal with it "before one's colleagues and competitors and thousands of strangers are reading all about it in the newspaper."[10] Morality seems not to have been a factor for either the administration or the newspaper, but in the end Plumb is correct when he says that in handling public power integrity is all, where "public image and private behavior must never diverge."[11]

The third incident that illustrates the dilemma over personal privacy in a public setting is the one that involved Vanessa Williams, who was dethroned as Miss America in 1984 because of nude photographs of her published in *Penthouse* magazine. *Penthouse* was protected by the First Amendment against any privacy claim, even if the pictures had been used without Miss Williams's consent. Although most people were said to have great sympathy for the beauty queen's misfortune and little for the magazine's claims of public service, they also showed great interest in seeing the published photos.

Richard Kurnit, writing in the *New York Law Journal*, suggested the inevitable paradox created when such privacy issues arise: "At some point, these photographs became a matter of public interest entitled to First Amendment protection. The question may well be whether that was only after *Penthouse* published them."[12] The media have consistently pleaded "newsworthiness" as their defense in private-facts cases. But the courts have noted on occasion that, because the press decides what is news, plaintiffs who are not public officials or public figures may find themselves in an intolerable double-bind with the Constitution. With public officials or public figures the situation is entirely different, as we have seen, for their private behavior is beyond the reach of legal protection.

Miss Williams's encounter took place in New York, where the right to sue for an invasion of privacy is extremely limited. The New York Court of Appeals, the state's highest adjudicator, has always taken the position that there exists no so-called common law right to privacy. The only claims are those provided in the New York Civil Rights law, passed in 1903 as the nation's first privacy statute and discussed in Chapter 2.

Over the years the New York court has limited the scope of the statute to its original intention: allowing for a cause of action by a person whose name or picture is used without prior written consent in an advertisement for a product. The law encompasses only the commercial use of an individual's name or likeness and no more.[13] Many federal court decisions and others in lower courts have tended to expand on the right of privacy and the right of publicity, but in New York the statute means what it says. Television actress Brooke Shields learned this when she failed in her suit over the publication of nude photographs taken when she was ten years old because her mother had signed an unlimited release.

Whether Miss Williams authorized the use or publication of the photos, her privacy claim is precluded by the First Amendment. Regularly the courts have ruled that "a picture illustrating an article on a matter of public interest is not considered use for purposes of trade or advertising . . . unless the article is an advertisement in disguise." This "holds true though the dissemination of news and views is carried on for a profit or that illustrations are added for the very purpose of encouraging sales of the publication."[14]

In *Flores v. Mosler Safe Co.* (1959), a New York federal court determined that "advertising purposes" means use in a solicitation for patronage or to promote the sale of some collateral product or service. When Joe Namath, the professional quarterback, sued *Sports Illustrated* for invasion of privacy under the New York Civil Rights law, the court established that there is a First Amendment right to use the name or picture of persons discussed or portrayed in "truthful" advertising. Further, the photos were newsworthy and used in the promotional ads to establish the news content of the magazine.

When the U.S. Supreme Court ruled the first time in a suit involving privacy and the press, *Time Inc. v. Hill* (1967), it stated:

Exposure of the self to others in varying degrees is a concomitant of life in a civilized community. The risk of this exposure is an essential incident of life in a society which places a primary value on freedom of speech and of press.[15]

As Kurnit said in his analysis of the Vanessa Williams case, the result is sometimes unpleasant and unfair to the individual, but for better or for worse, "we rely on the taste and ethics of the media and of the consumers who purchase or avoid certain publications, rather than on the government or on judges, to determine what is properly of interest to the public."[16]

This philosophy is the one that has flowed most frequently from the significant privacy-press cases, but the court rulings appear to have been based as well on a different philosophy— the public's right to know information, rather than the press's right to disseminate information. Judge J. Skelly Wright explains the confusion engendered by this change in philosophy:

At one extreme is what might generously be called the public's concern or curiosity about the private affairs of private individuals. At the other extreme is the public's interest in knowing the qualifications and views of candidates for public office and the considerations underlying passage or defeat of proposed legislation. At some point along this spectrum society's interest acquires legitimacy and at some point further that interest might be said to assume the status of a right.[17]

If news is by definition a matter of public interest, then the press must be the final arbiter of what is of public interest. With the showing of the existence of public interest left to the press, therefore, it would be difficult for a plaintiff to claim an invasion of privacy once the information had been disseminated. It could be argued that this shift in direction, if not in elementary philosophy, is ill-founded and that a public society is better served by the rights it bestows on its speakers. Citizen rights *to* information are derived *from* the citizen's right to speak.

Claims against the media for privacy invasion fall into the four torts identified by Dean Prosser and presented in Chapter 3. They are: intrusion, public disclosure of truthful private facts (what Warren and Brandeis sought to protect), false light in the public eye, and the commercialization of or the appropriation of name or likeness. Intrusion is broader than trespass and may include the use of electronic devices or photography from a distance. Private facts depend upon whether the public needs to know or is entitled to know the information. False light includes fictionalization, or the embellishment of an otherwise true story with little falsehoods, and the publication or broadcast of false information that creates a false impression. Appropriation (more accurately, misappropriation), the oldest and easiest to understand of the invasions, applies to advertisements for commercial gain without the person's consent. The previous chapter dealt with the fourth tort, from which has emerged the right of publicity/personality.

The *Restatement (Second) of Torts* defines the elements of an intrusion claim as follows:

> One who intentionally intrudes, physically or otherwise, upon the solitude or seclusion of another or his private affairs or concerns, is subject to liability to the other for invasion of his privacy, if the intrusion would be highly offensive to a reasonable person.[18]

Intrusion usually affects the news-gathering process, rather than the publication or broadcast of the information. The tort is based upon the overt acts of the intruder and not upon the subsequent publication or broadcast of information. However, not all courts have made this essential distinction. Intrusion, which is derived from common law trespass, includes surreptitious surveillance and instances where reporters exceed the limits of the consent to enter private property. *Dietemann v. Time Inc.* (1971) is the classic case.

A. A. Dietemann, a disabled veteran and journeyman plumber in California, practiced medicine in his home using clay, minerals, herbs, and gadgetry. Two *Life* magazine reporters, working with police, pretended to be married when they visited Dietemann's private dwelling. During the examination of one reporter, the other took photographs with a hidden camera. The reporters, who never revealed their true identities, also taped the conversation with Dietemann through a hidden microphone. After Dietemann was arrested, the magazine published a story on medical quackery and included the photos of the "doctor" taken by the hidden camera. In the intrusion suit, the court, while agreeing that the photos were newsworthy and *Life* had a constitutional right to publish them, ruled nevertheless that invasion had taken place with the secret recording and photographing of Dietemann in his home office. Judge Shirley Hufstedler said:

> Investigative reporting is an ancient art; its successful practice long antecedes the invention of miniature cameras and electronic devices. The First Amendment has never been construed to accord newsmen immunity from torts or crimes committed during the course of news gathering.[19]

Later on, when a public celebrity's privacy was ruled invaded by an American *paparazzo*, the court said that such behavior was not protected by the First Amendment but constituted assault, battery, harassment, and in violation of New York's privacy statute. But Ronald Galella, an aggressive photographer, argued that Jacqueline Kennedy Onassis's suit interfered with his right to make a living, but the court rejected his claim. Hidden, secretive devices may be used, however, if activities of public officials, such as policemen, are being recorded or if the undercover journalistic reporting occurs in a public place.[20]

Illegal intrusion, even when the product may result in a public-interest news account or a police arrest, is seldom seen as reason enough to invade personal privacy. As Don R. Pember points out:

> Defending illegal behavior on the grounds that it is in the public interest is an excuse which the American people seemed to reject

when the members of the Nixon White House used it in 1973 and 1974. Intrusion as an aspect of invasion of privacy should really not be a problem to journalists who conduct their business in an ethical fashion.[21]

The private-facts tort is far more controversial and has few guidelines to influence journalistic behavior. This is one area in which absolute, complete, and unmitigated truth will not constitute an ironclad defense for the media. In *Cox Broadcasting v. Cohn* (1975),[22] the Supreme Court made it clear that truth may never be an absolute defense for the media when they reveal the private life of an individual. A state may not impose civil liability based upon the publication of a rape victim's name because it was obtained from the public record, the Court said. But it decided not to address the broader issue of whether truthful publications may ever be subjected to liability.

In that case, the identity of a deceased rape victim was obtained by a reporter from court records—which are normally available to the public—and later her name was broadcasted. The victim's father sued the newsman and the television station for invasion of privacy, citing a Georgia criminal statute making it a misdemeanor to publish or broadcast the identity of a rape victim. The Georgia Supreme Court, declaring that a rape victim's name was not a matter of public concern, held that the statute was a legitimate constitutional limitation on the press's First Amendment right. The U.S. Supreme Court reversed, holding that the state may not impose sanctions "on the accurate publication of the name of a rape victim obtained from public records—more specifically, from judicial records which are maintained in connection with a public prosecution and which themselves are open to public inspection."

Cox is important for other reasons. First, in terms of privacy protection, the publication of the name of a *dead* rape victim seems inappropriate because that information would no longer matter socially or economically. The publicity caused distress to the victim's family and provided no useful information. Thus, the state court acted properly in holding that the broadcast invaded her family's privacy, which could have been avoided at no risk to either party had the journalist simply reported the story without the victim's name.

Second, in terms of press protection, the narrowness of the Supreme Court's decision becomes apparent when one realizes the limited nature of information found in public records. Most are of police and judicial proceedings, normally available to the public with little effort. For, as Gerald G. Ashdown properly notes, public records represent a relatively small portion of the total amount of information capable of producing embarrassing and undesired publicity. Arguably, the citizenry needs much more information than that required by law to be kept if it is to govern itself successfully.[24]

If, or when, the high court decides to confront the broader issue, it will probably apply some standard of negligence as it has done in defamation rulings, establishing a lower standard of fault for private persons than for public-official and public-figure plaintiffs. Justice White's language in *Cox* is supportive of privacy in this broader context and may portend the future:

> Powerful arguments can be made, and have been made, that however it may be ultimately defined, there *is* a zone of privacy surrounding every individual, a zone within which the State may protect him from intrusion by the press, with all its attendant publicity.[25]

In another oft-cited private-facts case, *Sidis v. F-R Publishing Co.* (1940) discussed in Chapter 2, the name of the individual whose privacy had been invaded was unavoidably part of the story. Had the *New Yorker* chosen to conceal Sidis's real identity effectively, it would also have had to alter other details, and the changes would have reduced the value of the story and raised doubts in readers' minds as to whether they were reading fact or fiction. This is certainly one way for the media to avoid privacy invasion suits, or libel, too, for that matter, but at high risk to their integrity and that of the First Amendment. The Sidis story was newsworthy in that it enhanced the public's interest in child prodigies—especially in a mathematical genius who was graduated from Harvard at age sixteen and was later found to have rejected the American dream of success! With the story of the dead rape victim, however, the public's interest in that kind of crime would have been served just as well with but one fact missing.

The *Cox* decision elicited other comments from Justice White, who, in referring to the Warren and Brandeis article, noted that "the century has experienced a strong tide running in favor of the so-called right to privacy," and suggested that there were "impressive credentials" for such a right. He said that previous cases, pointedly *Time Inc. v. Hill,* had "expressly saved the question whether truthful publication of very private matters unrelated to public affairs could be constitutionally proscribed." In this the Court indicated its interest in privacy and hinted strongly that it might be inclined to honor such claims when the broad question of media liability for public disclosure of private facts is eventually confronted.

In other words, *Cox* was based on established precedent and settled doctrine, but future private-facts claims may not receive the same pro-media consideration. The Court has left the door ajar. How this eventuality will affect the American tradition of free expression and public openness is a matter of conjecture. But what could be alarming is if public officials and public figures sought cover under the more flexible privacy tort than under the less protective shield of libel law.

Privacy law has a way of limiting access to information and can be an effective method for restraining the media prior to publication, as the efforts of Elizabeth Taylor and Frank Sinatra illustrate. Libel law, on the other hand, is only applicable after publication, when the burden of proof is more clearly the plaintiff's responsibility. More important, truth is the single best defense in a libel action; in privacy suits it may not come into play at all. The media believe that their best defense in private-facts disclosure is newsworthiness, as implied by the First Amendment, but the judiciary has been reluctant to interpret the Amendment that broadly.

Unless the media adhere to the *Cox* standard—the accurate publication of private information taken from public records—they may find themselves faced with potential liability in all other cases of unwanted public exposure. Information not obtained from public records may even fail to meet the test of self-governance or general political significance. "Given the Court's view of the appropriate balance to be struck between private rights and freedom of expression," Ashdown points out, "it is likely that *Cox Broadcasting* signifies the media side of the balance."[26]

Judge Cardamone's brief description of newsworthiness in the Jackie Collins case is also instructive. He said that the court could not accept a view that a publication must meet some independent standard of newsworthiness for it to stand under the umbrella of the First Amendment. Even "vulgar" publications are entitled to such guarantees, he said.

> It makes no difference that *Adelina* [the magazine in question] may have few redeeming features, that it may express a point of view far afield from what one might consider the community's standard of decency, or that an ordinary reader may find it distasteful. The compass of the First Amendment covers a vast spectrum of tastes, views, ideas and expressions. To hold otherwise would draw a tight noose around the throat of public discussion choking off media First Amendment rights.

Factual errors do not alter the subject matter of offending publications, the judge said, nor are they grounds for liability, absent proof of fault. Courts are hesitant to try to define what is news; the Supreme Court in *Gertz* specifically warned against "committing this task to the conscience of judges."[27]

The appropriation tort, as defined by Dean Prosser and which treats the commercial aspects of personal privacy, was the topic of the last chapter as it applies to the new right of publicity/personality. Not much more needs to be said, except to note that some courts have extended the right of publicity to cover more than just a celebrity's name and likeness. Other aspects of an individual's identity and public persona may also be protected. An interesting case is the 1983 federal appellate ruling in *Carson v. Here's Johnny*.[28]

Entertainer Johnny Carson, host of television's "Tonight Show," was not amused when the firm, Here's Johnny Portable Toilets Inc., called its product "Here's Johnny" and added the phrase, "The world's foremost commodian." After the trial judge dismissed Carson's suit on the violation of his right to publicity, the appellate court held that the use of "Here's Johnny" as a brand name did violate the entertainer's same right. A person's full name need not be used for his right of publicity to be violated, especially in a case involving a celebrity as well-known as Johnny Carson. The court noted the company's admission that it had tried to capitalize on Carson's reputation.

However, although catch-phrases that are closely associated with celebrities may be off-limits to advertisers, the Constitution protects the news reporting of such slogans. The right of publicity protects celebrities and others from the unauthorized commercial use of anything that may be identified with their personalities, but news is not considered to be commercial in the same sense. Yet, given the judiciary's current proclivity toward private rights over those of the public, it may be only a matter of time before "news" of public persons is classified according to a monetary standard instead of by the public interest test.

Regarding false light, in *Time Inc. v. Hill*, the classic case, the Supreme Court substantially protected the press from privacy actions regarding the publication of untrue but nondefamatory information. If the statements about the plaintiff are both false and disparaging, the correct route is libel and not the privacy action of falsification. "The plaintiff should not be able to skirt the requirements for defamation by giving his action a 'privacy' label," writes Ashdown.[29]

The case involved the James J. Hill family, which in 1952 had been held hostage in their suburban Philadelphia home by three escaped convicts, two of whom were later killed in a shoot-out with police. Media attention soon died down, but was revived less than a year later when novelist Joseph Hayes published *The Desperate Hours,* a story remarkably similar to the Hills' experience. The book was eventually made into a successful Broadway play and a Hollywood film. It was the play, as reviewed by *Life* magazine, that incited the legal action.

James Hill sued for invasion of privacy after the appearance of the magazine article, which included photos taken of the Philadelphia tryout cast in front of the Hills' former home. The headline read, "True Crime Inspires Tense Play," and the story called the drama a reenactment of the Hills' ordeal. Hill sought damages on grounds that the magazine article was inaccurate and constituted fictionalization, as forbidden under the New York privacy statute. He also said that *Life* had used the family's name for trade purposes, also illegal without consent, and that the article put the family in a false light.

After the Hills won a $30,000 judgment in the New York state courts, Time, Inc., publisher of *Life,* appealed to the U.S. Supreme Court and won a reversal in 1967. Freedom of the press became central, for the *Life* story was said to contain only one error—that the Hill captivity alone had inspired the book and the play. Novelist Hayes, who also wrote the play, had been dropped from the suit when he made it clear that his works were based upon other incidents as well. Previous cases won by defendants and deserving the label fictionalization had contained many more errors reflecting poorly upon the plaintiff.

In *Hill,* the court ruled that a plaintiff was not entitled to "redress [for] false reports of matters of public interest" unless it could be shown, as in libel law, "that the defendant published the report with knowledge of its falsity or in reckless disregard of the truth."[30] Carefully emphasizing that it was limiting its application of the *New York Times v. Sullivan* doctrine to the false-light category of privacy, the Court accomplished two broad results in its decision: 1) it strengthened the media's defenses against invasion of privacy suits; and 2) it blurred the distinction between privacy interests and reputation interests. On the latter result, Melville B. Nimmer argues that "the Court fell into error by reason of its failure to pierce the superficial similarity between false-light invasion of privacy and defamation" and also by reason of its failure to create a doctrine that would recognize the particular interests of the right of privacy. Nimmer questions the media's First Amendment privilege in privacy matters, recommending the complete denial where false-light invasion is an extension of public disclosure of embarrassing private facts, or, that is, when no injury to the reputation has occurred. Because the libel standards of "public interest" and "reckless disregard" have also become decisive factors in privacy suits against the media, the recent history of those concepts is important to recall.[31]

As it turns out, the most important privacy case is the landmark libel case *New York Times v. Sullivan* (1964), which began when an elected commissioner of Montgomery, Alabama, who supervised the police department, sued the newspaper for publication of a paid advertisement that accused the city of mistreating black students protesting against segregation. The plaintiff, L. B. Sullivan, won in the Alabama courts, but the Supreme Court ruled that the state law did not allow enough leeway to the press in writing about public officials. As Justice Hugo Black said in a forceful concurring opinion, "An unconditional right to say what one pleases about public affairs is what I consider to be the minimum guarantee of the First Amendment." This logic, as we have seen in *Hill*, for example, has been applied to public-interest privacy cases as well.

After *Sullivan* and subsequent decisions, culminating in *Gertz v. Robert Welch* a decade later, the plaintiff had to show "actual malice," or "reckless disregard" of the truth, to prevail in litigation brought by public officials and public figures. With *Gertz*, even private plaintiffs had to show some degree of fault, at least in press cases, but the court left the standard of fault to the discretion of each state. In New Jersey, for example, the standard is "negligence" for private individuals and "actual malice" for public persons.

Elmer Gertz, a prominent Chicago attorney and law professor, slowed down the post-*Sullivan* mood of the court to favor media defendants when he convinced the court that, despite his sometimes public prominence, he was acting as a private person in representing the family of a young black man who had been shot and killed by a Chicago policeman. The court believed that Gertz had done nothing to seek public figure status in this case and, thus, should not be required to prove actual malice to win his libel suit against the publisher of the John Birch Society's magazine *American Opinion*. The article in question had falsely accused Gertz of various subversive activities and called him a "communist-fronter" and "Leninist."

But the Supreme Court continued to adhere to its basic and continuing rationale for open and robust debate on issues of public concern. It reiterated that position in *Gertz*:

> Our decisions recognize that the rule of strict liability that compels a publisher or broadcaster to guarantee the accuracy of his factual assertions may lead to intolerable self-censorship. Allowing the media to avoid liability only by proving the truth of all injurious statements does not accord adequate protection to First Amendment liberties.

Constitutional privilege became a defense against libel, affording a high degree of immunity to the media even though the offended individual may have been defamed, but not intentionally. When the privilege is applied, society's interest in open public debate is considered to outweigh the individual's interest in her or his reputation. The privilege may be absolute, as it is in civil libel in New Jersey, or conditional, as when, in criminal cases, the Sixth Amendment guarantee may be held more important on balance than the First Amendment.

The court held that misstatements of fact or even unjustified or unfair comments or opinion about the conduct of public officials were privileged unless the information was published with "actual knowledge of falsity or with reckless disregard of probable falsity." By submitting public officials to a tougher standard, the burden of proving the material false shifted to the plaintiff. The old strict liability rule of common law presumed falsity upon publication.

Behind the philosophy that "debate on public issues should

be uninhibited, robust, and wide open" is the belief that most public-official and public-figure suits are really sedition cases designed to punish the news media for criticism of government affairs. Seditious libel nearly always means prior restraint. Furthermore, there is the suspicion that these kinds of lawsuits tend to quell debate on political issues in general. Public officials must expect to be criticized in public, sometimes quite severely and unfairly—it goes with the territory. The recent cases of Generals William C. Westmoreland and Ariel Sharon come to mind.

In the suit brought by Sharon, Israel's former Defense Minister, the jury found that a 1983 *Time* magazine article, which contained a false and defamatory paragraph, did not libel Sharon and concluded that the magazine had acted in good faith. *Time* had asserted in its interpretation of the official Israeli government report on the Sabra and Shatila massacres by Christian Phalangists that General Sharon had been more than indirectly responsible for the killings, which took place during the Israeli invasion of Lebanon in the fall of 1982 and immediately after the assassination of Lebanese President-elect Bashir Gemayel. The paragraph in question, which *Time* could not verify, enraged Sharon, who accused the magazine of having committed "blood libel," an historic anti-Semitic device.

We will never know what the jury might have found in the Westmoreland case, for near the end of the trial the general withdrew his suit against CBS. He had charged that the network had libeled him and his staff by accusing them of having participated in a "conspiracy" to show Washington that progress was being made in the Vietnam war by deliberately underestimating the size and capabilities of the enemy. Although CBS was not able to prove beyond doubt the truthfulness of its allegations, the general was said to have been so demoralized by damaging testimony during the trial that he believed his suit a "no-win situation." Westmoreland, a public figure like Sharon, suspected he could not show actual malice, or reckless disregard, in the CBS documentary. In both cases, the fearful element of sedi-

tion was apparent, even though the media may have been care-
less and negligent in their journalistic assessments. The Sharon
jury, in an unusual move, publicly criticized *Time* at the same
time it announced its finding of no malicious libel.[32]

After *Sullivan,* the court extended its new standard to public
figures as well and said that public interest was also a determining element. Existence of the constitutional privilege should depend upon the newsworthiness of the event and not upon the
public status of the individual involved. The court extended
First Amendment protection to media reporting of matters of
public or general interest even if a private person were
involved.[33]

The media rejoiced in each expansion of the First Amendment, but their brief absolutist tradition came to an abrupt halt
with *Gertz.* "Sensing that the balance between free speech and
private reputation had tipped too far in the direction of free
speech," the Court retreated to a limited extent and invited the
states to fashion a similar retreat. It said that the states "should
retain substantial latitude in their efforts to enforce a legal remedy for defamatory falsehood injurious to the reputation of a
private individual." But the Court held to its malice standard for
public officials and persons who are clearly public figures, and
said that the standard applies to any plaintiff, public or private,
who desires punitive damages.

In rejecting the media's argument that Elmer Gertz, a prominent Chicago attorney, was a public figure, the Court said that
only those individuals who *voluntarily inject* themselves into a
particular public controversy are considered limited-purpose
public figures who must overcome the malice test. The Court, in
a subsequent case, emphasized that the higher standard did not
hinge on whether the statement concerns a matter of public
interest. A celebrated divorce involving a prominent and
wealthy couple was not deemed a "public controversy." Such a
plaintiff would only have to show simple negligence in a private-person libel suit. A federal circuit court later defined the
"essential element" underlying the new category of public fig-

ure as when "the publicized person has taken an affirmative step to attract public attention."[34]

An important question in each of these cases is what is "the nature and extent of an individual's participation in the particular controversy giving rise to the defamation" and to personal privacy, as *Gertz* replaces *Sullivan*. If *Hill* were on the docket today, the Court would apply its new formulation, shifting the focus of constitutional privilege away from public interest and back to the character of the individual and finding for James Hill, who could no doubt meet the negligence standard for private persons. As Marilyn A. Lashner writes,

> . . . it seems reasonable to expect that the emerging liability of the press would accompany an expanding protection from privacy invasions—a protection where false-light non-reputation-injuring publications would be recognized as an extension of disclosure of private facts, and where both are subsumed in a single doctrine.[35]

Such a change in focus may be in keeping with America's upsurge in privacy interests, but it does not bode well for the media, which traditionally have served the public interest. And, when coupled with the Court's deceptive "right to know" concept, the community, as well as the media, may have real cause for concern. At this juncture it should be apparent that the media's watchdog role is at once ironic and contradictory. On the surface, "right to know" appears to serve the public as a positive response to the consumers of information. Because Americans want it both ways—freedom of expression *and* personal privacy—their courts and legislatures have interpreted the Constitution to embrace the rights of media consumers. In so doing, however, as Gerald J. Baldesty and Roger A. Simpson perceptively warn, the Supreme Court has elevated the idea of a right to know to such an extent that the traditional imperative of a right to speak can no longer be confidently assumed. "Hollow rights have been advanced on behalf of consumers to justify governmental controls on press content," they write.[36]

The contradiction in privacy law is that, whereas government interests are the same as community interests, the fact is that when the state decides to protect personal privacy it does so at the expense of that very privilege. For, as Associate Justice Brennan stressed in *Sullivan,* the "protection of the public" requires full discussion and complete information. If basic free speech is guaranteed, it follows that the rights of information consumers are protected. The irony is that if the courts were to broaden their definition of "public," as appeared evident before *Gertz,* the individual's private interest would be better served and thus more in line with a sense of community. The double jeopardy nature of "right to know" becomes clear when we admit to the fallacy of ears and eyes determining what tongues shall speak. As Associate Justice Brandeis said in 1927, the freedom "to think as you will and to speak as you think" is essential to the search for truth.[37] To put it another way, the rights of the individual are derivative of the rights of the public.

Like a juggler, the U.S. Supreme Court has managed so far to secure equivalent protection for reputation, privacy, and the media. Because the three rights are so intertwined, the Court has had to deal with them collectively in decisions involving privacy and the press. In its first such encounter, *Time Inc. v. Hill,* the justices gave preferred status to the press. In subsequent decisions, however, a majority of the Court has emphasized the importance of protecting both freedom of expression and privacy. Without completely merging privacy and defamation, the Court has developed parallel reasoning in these separate areas of law. But to date, no broad conceptual base has emerged. For example, in the same term that they decided *Cantrell v. Forest City Publishing Co.* (1974) and *Cox,* the justices declined to decide another case that, unlike those cases, appeared to raise some of the general issues surrounding the constitutionality of relief for true statements invading privacy.

In *Doe v. Roe,* a former psychiatric patient sued to restrain distribution of a book disclosing "confidential communications" between the patient and her psychiatrist. She maintained that publication of "the near verbatim record of her psychotherapeutic treatment" constituted a breach of confidence and an invasion of her right of privacy. The trial court granted a preliminary injunction against distribution of the book, and the New York Court of Appeals rejected the argument that such prohibition would "constitute an invalid prior restraint upon publication." *Cantrell* and *Cox*—the former favoring the plaintiff and the latter supporting the press—held that truthful publications were actionable despite the public interest nature of the news stories. *Cox* hinged on the very narrow grounds of the source—the public criminal record.[38]

"The community's notions of decency," a factor in *Hill*, was used in 1974 by a District of Columbia trial court to rule that the publication of the name of a twenty-one-year-old rape victim was not constitutionally protected. contrary to the approach taken in *Cox*. The court said that to warrant protection, a truthful statement must concern a matter of public interest and must not shock the community's sense of decency. The Pennsylvania Supreme Court held that a truthful disclosure of the plaintiff's debt to his employer and three of his relatives did not constitute an invasion, but indicated that such a disclosure to the public at large would be actionable because the gravamen of the action is "publicity."[39] Even *Zacchini*, whose significance was discussed in Chapter 7, is not much help in the formulation of a conceptual base for privacy cases involving the media. It was decided on limited grounds.

The First Amendment has long protected the media's role as disseminators of information, but it has not always supported the media in their role as information gatherers, upon which dissemination obviously and ultimately depends. What journalists acquire they are normally free to publish or broadcast, although gaining access to information has not always been easy or guaranteed by law. With privacy, the personal right has become an impediment to news gathering under a variety of laws and regulations. For many years, the Sixth Amendment, one of several due process provisions in the Constitution, was the primary argument against access to proceedings in criminal matters. But now privacy interests are beginning to displace Sixth Amendment concerns. Recently, the Supreme Court held that the First Amendment and the privacy rights of members of a religion to keep their membership and financial contributions secret outweighed the media's right of access to such information.[40] The Court limited access to information obtained during pre-trial discovery, for, in an earlier decision, the justices said that "the presumption of openness may be overcome only by an overriding interest based on findings that closure is essential to preserve higher values."[41]

Much of the difficulty with access has occurred in pre-trial and post-trial criminal proceedings—the court reaffirmed public access to the actual trial in *Richmond Newspapers v. Virginia* (1980). In *Press-Enterprise v. Superior Court* (1984), a California case, the newspaper sought release of the transcript of approximately six weeks of *voir dire* proceedings—the process whereby prospective jurors are questioned to discover bias before they are empaneled. The trial involved the rape and murder of a teenage girl, similar to the Cox case. Prosecution objected to the paper's access to the proceedings, claiming that this could inhibit the candor of the jurors' responses to questions and thus jeopardize the defendant's Sixth Amendment rights. The court permitted press access to the general *voir dire*, but excluded the media and the public from the special death penalty questioning.

After the jury had been empaneled, the *Press-Enterprise* moved for release of the *voir dire* transcript, which the state and the defendant opposed on the grounds that such exposure would violate the jurors' privacy rights because they had been questioned under a promise of confidentiality. The trial court denied the newspaper's motion, but a unanimous U.S. Supreme Court ruled that the decision had violated the right of the public and the press to attend criminal proceedings. Although the opinion relied on First Amendment access cases, the only specific reference to freedom of the press is in a footnote: ". . . the question we address—whether the *voir dire* process must be open—focuses on First, rather than Fifth Amendment values and the historical backdrop against which the First Amendment was enacted."[42]

Press-Enterprise recommended a standard (albeit vague) for closure: "Closed proceedings, although not absolutely precluded, must be rare and only for cause shown that outweighs the value of openness."[43] On the issue of juror privacy, the court left the door open and implied that in a different case a juror's right may extend to the *voir dire* process. A recent Practising Law Institute (PLI) seminar paper noted that, ". . . whether the access right extends to a particular proceeding or specific record depends . . . on whether access would secure to the public information relevant to the discussion of governmental activity."[44] All aspects of the criminal justice system would appear to qualify as relevant to the body politic. In access claims, the public's right to know information that is related to self-governance appears now to carry weight nearly equal to the private person's right to privacy.

But the PLI seminar paper also alluded to another recent media case that extends the recognition of a privacy interest to another area. In *Tavoulareas v. Washington Post*,[45] a panel of the District of Columbia Circuit Court affirmed a gag order prohibiting the newspaper from disseminating information it had obtained during the discovery proceedings as a defendant in a libel suit. In effect, the panel created a general right to privacy for corporations. The panel said that any statutory presumption of openness was outweighed by a corporation's constitutional right to privacy. Access to anything other than documents used at the trial itself is limited by federal law, which permits gag orders for trade secrets. Corporations have a right of "disclosural privacy" with respect to commercial information, the court said. That is a reasonable position, particularly in that it does permit public access to trial information. It limits the media's ability to report on the internal workings of the so-called private sector, which, history teaches, bears watching from time to time.

Each of the cases discussed in this chapter has contributed in one way or another to the emergence of a broad constitutional protection for the media, at the same time allowing for some degree of privacy to individuals. Where there is a clear indication of public interest, whether interpreted as such by the judiciary or the media, decisions usually have favored the press. However, there persists a good deal of ambivalence as to how far the right of access may be pushed, given the tension that exists between public and private rights.

When the Supreme Court refused in 1965 to validate passports for reporters traveling to Cuba, it said, in typical paradoxical language, that "the right to speak and publish does not carry with it the unrestrained right to gather information."[46] And when the Court denied newsmen special access to prison inmates, the justices held that "newsmen have no constitutional right of access to prisons or their inmates beyond that afforded the general public."[47] These are but a few of the recent attempts at adjudication by consensus, America's way of dealing with its many different societal interests. Or, as the California Supreme Court commented, "The right to know and the right to have others not know are, simplistically considered, irreconcilable."[48] Meanwhile, the media may have to argue more forcefully—especially before the court of public opinion—that they serve the public interest.

9

Privatizing Information

Although the Supreme Court has never ruled explicitly on whether the public disclosure of embarrassing private facts as a cause of action is a permissible vehicle for the protection of personal privacy, it has recognized a constitutional right to privacy and has intimated that legislatures may recognize rights to privacy not grounded in the Constitution. Federal and state governments, in response to that implied mandate, have passed statutes with both closure and disclosure in mind. However, their response has not been directed solely at the mass media.

Many of the most troubling privacy questions today arise not from the media's wide dissemination of private information, but from the rise of technology that allows for the exchange of computerized information and the development of so-called databanks. Most citizens openly submit to a subtle invasion in order to obtain credit, medical care, or insurance, but the harm to privacy may be greater than mass publicity by the media. As Diane L. Zimmerman posits the issue:

> Privacy law might be more just and effective if it were to focus on identifying (preferably by statute) those exchanges of information that warrant protection at their point of origin, rather than continuing its current, capricious course of imposing liability only if the material is ultimately disseminated to the public at large.[1]

Computerized information stored in countless databanks may cause serious harm to reputation, even though its circulation is only to a couple of unauthorized recipients. Therefore, in Zimmerman's view, ". . . thoughtful elaboration of privacy law involving intrusions on solitude is likely to promote greater protection of the individual's interest in being free of public scrutiny than is the vague and hard-to-apply law governing the publicity of private facts."[2] Intrusion, not private facts or false-light publicity, is the focal point of modern technology as a means of invading privacy. And that is what the federal legislature set out to curb with a series of laws in the 1960s and 1970s. This chapter and the next examine the two main statutes that most directly affect openness in government and privacy of the individual— the Freedom of Information Act and the Privacy Act.

Professor Arthur R. Miller, whose book *The Assault on Privacy: Computers, Data Banks, and Dossiers* (1971), an early comprehensive look at the threat of computer-driven intrusion, identifies four recent developments that relate to the late twentieth-century concern for privacy: 1) massive record-keeping; 2) decision-making by dossier; 3) unrestricted transfer of information from one context to another; and 4) surveillance conduct at one level or another. "To me," he writes, "the modern concept of privacy does not relate to intrusion, misappropriation, embarrassing private facts, or false light. Those things constitute the work of lawyers in their quest to get things within rigid limits. I know an embarrassing fact and misappropriation when I see them."[3] But record-keeping and data-collection represent new and different ways of disrupting solitude and seclusion as more and more institutions collect more and more information about Americans and about more aspects of their lives.

Once collected, the information is used as the basis for making decisions. "The truth is that insurance, employment, credit, benefit eligibility, even admission to a university, are transacted almost exclusively on the basis of information, not person-to-person encounters," according to Miller. This growing dependence creates problems, even anxiety, as to the accuracy of the information, its currency, its relevance, and, in Miller's acerbic view, "the wisdom of the middle-level bureaucrats who manipulate and make decisions based on the information. Do we have any existence other than these files?"[4]

There is also the growing concern that the information collected is being used out of context, or, what is worse, used for reasons other than for what it was acquired and stored in the first place. For example, the FBI's Criminal Offender Record Information System, which is available to law enforcement agencies across the country, contains arrest information on people, many of whom will never be prosecuted and are therefore, under our system of justice, not criminals. But, as Professor Miller notes, they are called "criminal offenders" because their names are put in the arrest file as such. "Simply having one's name in that file is a stigma," he says. The problem is increased by the fact that in some states arrest records circulate beyond the law enforcement community and are used by licensing agencies as well. Thus, as Miller explains the danger, a system that was generated for law enforcement purposes is suddenly and curiously deemed relevant to the question of whether you can cut hair in Florida, sell securities in New York, or drive a taxicab in Denver.[5]

The clash between personal privacy and public openness is nowhere more apparent than in the two major pieces of legislation under review here. Both were heatedly debated in Congress before eventual passage. The Freedom of Information Act (FOIA), first enacted in 1966 and amended in 1974 and 1976, requires disclosure to the public of certain information collected and held by the federal government. The Privacy Act (PA), its counterpart enacted in 1974 in the same Congressional session that revised the FOIA, is meant to safeguard personal privacy in an effort to curb the government's use of its vast files of information on citizens. But because the policies of privacy and disclosure are, on the face, mutually exclusive, neither law has fully succeeded.

The PA stipulates that records that are required to be open

under the FOIA are not subject to the provisions of the Privacy Act. Frank Rosenfeld explains the dilemma confronting agency bureaucrats:

> If they refuse to disclose the material they risk being sued by the party who requested the file under the Freedom of Information Act. Under the FOIA the court may award to a successful plaintiff his costs and attorney's fees. If, on the other hand, agencies release material, they risk being sued under the Privacy Act by the person who is the subject of the file. In that case, the plaintiff might win by showing that the file was exempt from disclosure under FOIA. A successful Privacy Act plaintiff can collect not only his costs and attorney's fees but also actual damage sustained because of disclosure.

It is at this level that the first instance of the privatization of information occurs, because if the official is going to err in his decision it is less costly to withhold the requested information and risk a suit under the FOIA. In which case, the FOIA is a hindrance rather than a help, a contradiction, in the effort to open public records.[6]

Marc Arnold and Andrew Kissiloff have noted yet another conflict between the two federal laws. Prior to the enactment of the PA (which occurred less than a month after Congress overrode a presidential veto of seventeen amendments to the FOIA), any materials that were not required to be open under the FOIA were nevertheless disclosable at the discretion of the government agency. Now, as a practical matter, the two researchers point out, information falling under any of the FOIA several exemptions—meaning they are not required to be available—will invariably be withheld as a matter of agency preference, convenience, and habit, also out of fear of violating the Privacy Act.[7]

Ever since the two issues simultaneously achieved prominence during the early 1970s, tension has developed between the right of personal privacy and the right of access to information. Congress passed the FOIA to end bureaucratic secrecy and "to guarantee public access to Government information" in an effort to ensure the correct functioning of an informed electorate so "vital to the operation of a democracy." In the shadow of the

measure is the theory that the people can only operate effectively and assess their government critically if they possess the information policymakers had when making decisions. "A popular Government, without popular information, or the means of
acquiring it," James Madison wrote in 1822, "is but a Prologue
to a Farce or a Tragedy; or, perhaps both. Knowledge will forever govern ignorance: And a people who mean to be their own
Governors, must arm themselves with the power which knowledge gives."[8]

Toward that end, the FOIA requires government agencies to
make records available to any person upon request, except
when the material is specifically exempted by the statute. There
are nine categories of exemptions:

1. Records specifically authorized to be kept secret in the interest of national security;
2. Records related solely to the internal personnel rules and
   practices of an agency;
3. Records specifically exempted from disclosure by statute
   (e.g., Privacy Act);
4. Trade secrets and commercial or financial information obtained from a person and privileged or confidential;
5. Inter-agency or intra-agency memoranda or letters not normally available by law other than for litigation purposes;
6. Personnel and medical files and similar files, the disclosure
   of which would constitute a clearly unwarranted invasion
   of personal privacy;
7. Investigatory records compiled for law enforcement purposes;
8. Records contained in or related to examination, operating,
   or condition reports prepared by, on behalf of, or for the
   use of, an agency responsible for the regulation or supervision of financial institutions; and
9. Geological and geophysical information and data, including maps concerning wells.

Abbreviated, the FOIA exemptions cover national defense secrets, trade secrets, certain law enforcement records, and some inter- and intra-agency correspondence. However, an agency is not perforce obligated to withhold exempt documents, as the Supreme Court determined in *Chrysler Corp. v. Brown* (1979).[9] The holding was in part based on the fact that the FOIA gives federal district courts the power to compel disclosure of information not falling within one of the categories but gives no corresponding power to block the disclosure of exempt material.

Since the FOIA went into effect on July 4, 1967, both Congress and the courts have struggled against bureaucratic efforts to further privatize various agency records. (As used in this chapter, the terms "privatize" and "privatization" are variants of "private" as opposed to "public" as these are employed throughout the book. "Privatize" and "privatization" are also used to signal the periodic shift away from government-financed programs and toward private-sector contracts and sponsorship of otherwise public programs. However, in both contexts, the implied result is the same—the withholding of information from the public.)

More often than is necessary the FOIA has been used to limit public access through the application mainly of the privacy exemption. Kimera Maxwell and Roger Reinsch note succinctly in their recent study: "A variety of federal and Supreme Court cases show that, instead of protecting only the privacy of individuals, this exemption [Exemption 6] has been used by custodians in government agencies to protect the privacy of the agencies."[10] The agencies also abuse the rights of the individuals they profess to protect by citing this exemption, even though the individuals may have no objections to the release of the information, Maxwell and Reinsch assert. Most government agencies fought the FOIA from the beginning, creating broad definitions of exemptions and claiming that requested material could not be found. They also charged exhorbitant fees and delayed searches to discourage use of the act.

Costs of facilitating the act, no doubt higher than necessary or even beyond what Congress itself had anticipated, have been used by the Reagan administration as a reason to amend the FOIA and thus restrict the public's access. Attorney General William French Smith in May 1981 instructed agencies to provide data on the cost of implementation and other information about FOIA compliance. The Justice Department then used the information to support the Administration's case for amending the law. The department claimed that the cost of complying with the FOIA in 1980 was about $57 million, "hundreds of times more than Congress contemplated." Congress believed neither cost nor any of the other Administration problems with FOIA to be sufficient grounds for altering it. But the cost of running the statute, whether at expense to the government or to the applicant, remains a factor in the overall effectiveness of the FOIA.[11]

In 1974, Congress overrode President Ford's veto of a series of amendments designed to correct some of the agency abuses, and reaffirmed its commitment to "the efficient, prompt and full disclosure of information."[12] One amendment called for the release of "any reasonably segregable portion" of otherwise exempt records. Another narrowed the exemption for "investigatory records compiled for law enforcement purposes," while other revisions sought to strengthen sanctions for noncompliance in an effort to discourage agency delay. Congress also limited fees to the actual costs for searching and copying material.

The amendments required agencies to announce their request procedures, to expedite appeal guidelines, to adopt uniform search and duplicating costs, and to publish detailed indices of their holdings. Also, courts could rule that the government pay court costs and other legal fees for successful FOIA plaintiffs. Judges were empowered to review at their discretion government documents *in camera* to help them decide whether any of the exempt categories had been correctly applied. Congress wanted to discourage rulings that permitted the blanket classification of documents without judicial review. For instance, when Congresswoman Patsy Mink and colleagues sought the release of interdepartmental reports on the advisability of underground nuclear tests in the Aleutians, the Supreme Court upheld the exemptions related to national defense and intra- and inter-agency memos (Exemptions 1 and 5) on the grounds that no *in camera* inspection was called for in the FOIA. In a case involving the CIA, a federal court of appeals permitted a number of book deletions without judicial review to determine what parts of the classified material remained harmful to national security.

In 1976, Congress again adjusted the FOIA, this time to control the proliferation and scope of statutes that exempt information from disclosure, but it left the privacy clause relatively unchanged. Exemption 6 allows for the closure of records that constitute "a clearly unwarranted invasion of personal privacy." Although the exemption is an honest attempt to balance individual privacy and the public's access to information, the result has been instead to balance two kinds of privacy—the individual's and the government's. In case after case, agencies have tried to protect their own privacy and in the process have further privatized public knowledge. Basic from the start, according to some observers, is that bureaucrats know far better how to use the law to withhold information than citizens know (or perhaps care) how to use it to obtain information.

Steve Weinberg, in a review of the 1974 amendments published in *The Nation*, April 19, 1975, noted that the changes were substantive as well as procedural, but that even the substantive ones may have little effect. Of the nine original exemptions, only two have been substantially modified to make access easier—for classified documents and investigatory files. The others remained intact, allowing for these seven to be relied upon more frequently by officials determined to withhold information. And the one Weinberg most fears being used inappropriately is Exemption 5, which covers "inter-agency or intra-agency memorandums or letters which would not be available by law to a party other than another agency in litigation with the agency."[13]

The FOIA calls for the release of "agency records," and it applies to "records which have been in fact obtained," not information in the abstract, nor information that might be forthcoming from a source, nor records that merely could have been acquired. In *Forsham v. Harris* (1980), the court held that records of a federally funded university research project were *not* records subject to disclosure under the FOIA unless they had been taken over by a government agency for its own review or use.[14] In *Kissinger v. Reporters' Committee for Freedom of the Press* (1980), the court said that the FOIA does not require an agency to retrieve or create records.[15] It also raised the question of whether agency records transferred to a non-agency remain accessible under the law. The former Secretary of State had deeded some of his personal notes to repositories not subject to the FOIA (the Library of Congress, for instance, which is not an FOIA-defined "agency"), and the court supported the principle in both cases that physical custody did not always constitute "records" for purposes of the FOIA. But the Kissinger case is more than a matter of simple judicial interpretation of language; it is an episode in the movement to privatize public information.

While he was assistant to the President for national security affairs and later as Secretary of State in the Nixon administration, Henry Kissinger kept in his personal files notes and transcripts of his telephone calls. Shortly before leaving office, he transferred the files to the New York estate of Vice President Nelson Rockefeller and subsequently deeded the material to the Library of Congress, which is not subject to the FOIA. In the transfer agreement, Kissinger stipulated, as he could have with any such archive, that he would control public access to them for a specified period. In *Kissinger*, FOIA requests by the Military Audit Project, the Reporters' Committee, and *New York Times* columnist William Safire for copies of transcripts of the telephone conversations made by Kissinger during his time in the White House and as Secretary of State were turned down on appeal to the State Department.

A federal district court ordered the Library to return to the State Department transcripts relating to Kissinger's position as Secretary of State, because they were agency records subject to disclosure and had been removed without permission. Howev-

er, the court denied the requests for notes prepared while he
was national security adviser to the President. But the Supreme
Court, in an opinion written by Associate Justice Rehnquist,
denied access to all parties and said that courts may create remedies and enjoin government agencies only if an agency has im-
properly withheld agency records. Safire had sought the White
House adviser's notes, not State Department records. The Military Audit Project and the Reporters' Committee wanted records that were no longer in the control or custody of the State
Department, the "agency" in question, which was not obligated
to retrieve documents that it no longer possessed. What Safire
sought was in the possession of the State Department but outside of its control as material belonging to the President's personal staff and, therefore, not technically agency records subject
to the FOIA. As Donald M. Gillmor and Jerome A. Barron assess
the problem, "Possession without control was insufficient to
make the documents records for purposes of the act."[16]

Associate Justices Brennan and Stevens dissented in part because they disagreed with the majority's narrow definition of
"custody or control." Stevens noted that the ruling would encourage outgoing officials to remove damaging information
from their files. He said that an agency retains custody over
anything it has a legal right to possess. Gillmor and Barron
reported that others interpreted Rehnquist's opinion a reversal
of the presumption that the burden under the FOIA is on the
agency to prove that it was justified in withholding information
requested.

> It may be very difficult, as a threshold requirement, for an FOIA
> plaintiff to show that agency records were improperly withheld.
> And how does a requester prove that records, if indeed they were
> under agency control in the first place, are subject to the required
> degree of agency control?[17]

By making access to public records hinge on whether the requested material is in the "possession" of an agency, there is
great risk that such a bureaucratic technicality will mark the
return of government secrecy. Courts, whose decisions are nor-

mally based on subtle and narrow interpretations of law, may not be expected too frequently to address broader societal problems. Meanwhile, Marie Veronica O'Connell has suggested a new test for determining the status of agency records under the FOIA, especially those generated by so-called non-agencies, which are "public" in every other way.

The courts should recognize that "possession" may be broadly defined to include documents not physically within an agency but still under its control. "As the FOIA is aimed at disclosing the materials government uses," O'Connell writes, "where it is unclear who controls a given document, courts should adopt a functional analysis, looking to the use of the document to determine its agency record status."[18] So long as nonpossession of non-agency records is a legitimate way to avoid FOIA disclosure, it is likely that government officials will farm out sensitive business to outside agencies and transfer to non-agencies those documents they wish to shield from public scrutiny. "Ironically," O'Connell notes, "circumvention of the FOIA is most likely when disclosure is most necessary—when the contents of documents reflect poorly on an agency or its staff."

The FOIA was drafted originally to protect legitimate government interests in national security, business confidentiality, and law enforcement. But, as James Weighart, executive editor and former Washington bureau chief of the *New York Daily News*, testified before the Senate Judiciary Committee in 1981,

> It cannot be seriously claimed that the act has resulted in a compromise of those interests as a result of forced disclosure of information that should have remained confidential. Instead, what appears to be involved is a frontal attack on the principle of an open government that underlies the Freedom of Information Act.[19]

Since 1970, the Duke University *Law Journal* has published an annual review of the previous year's developments under the FOIA. There are several other organizations and individuals at the watch, but the Duke review is the best systematic look at the effects of FOIA litigation on public access. What follows are brief comments on the nine exemptions as analyzed by the journal's reviewers:

Exemption 1 is designed to prevent disclosure of properly classified records, the release of which would cause at least some "identifiable damage" to the national security, "specifically authorized under criteria established by an Executive order to be kept secret in the interest of national defense or foreign policy." In 1982, President Reagan exercised his power under the FOIA to set criteria for exempting information classified as secret and eliminated President Carter's order requiring officials to consider public interest in openness when deciding to classify documents.

The Reagan order also created additional categories of classification and permitted agencies to withhold information more easily. Furthermore, the order revised the Carter declassification standards, which had required a declassification date to be set at the time of classification—domestic information up for review every twenty years and foreign information every thirty years to determine if it was still necessary to classify the material as "secret." If the information was not reviewed at the end of the specified time, it was automatically declassified.

Under the Reagan order, mandatory review only occurs at the time the information is initially classified, or "born classified." Thus, as the Duke reviewers note, information not given a predetermined declassification date could be forever classified and exempt from public scrutiny. Moreover, the Reagan order eliminated the test of "identifiable damage" to the national security and the discretionary "public interest" balancing of the Carter order. It also represented a setback for the entire declassification process developed by the Carter administration.

Floyd Abrams, the prominent First Amendment attorney, asserts that many of the changes in the classification system under President Reagan are the product of anger by the intelligence community at the Carter administration. The Information Security Oversight Office (ISOO), which has the responsibility for the security of all executive-branch agencies involved with classified materials, explained that one reason for the rule changes was because those previously in effect sounded too "apologetic." Abrams points out that changes in the language between that of the Carter administration ("Information may not be considered for classification unless it concerns . . .") and that of the Reagan administration ("Information shall be considered for classification if it contains . . .") were justified as the substitution of "positive" words for "negative" ones.[20]

Despite the fact that Congress, in amending the FOIA in 1974, sought to overcome judicial deference to agency expertise in classifying information by permitting *in camera* review, judicial reviews have continued to respect an agency's interpretation of the classification system. The fox is still in charge of the henhouse. With executive orders and judicial reviews favoring the privatization of public information, it is difficult, if not impossible, for interest groups, let alone individual citizens, to prevail. When the singer Joan Baez lost her FOIA suit to obtain part of her Justice Department dossier, the court upheld the government's use of Exemption 1 because no agency "bad faith" could be shown and ordered her to pay the court costs of the Justice Department, because it was the prevailing party.

The Duke reviewers speculate that the White House may be reacting to a fear of excessive and dangerous leaks of sensitive security information. But the reviewers also note that such fear may be ill-founded, for at the time of their 1984 critique, only six leaks had been reported in three years to the ISOO. Abrams accuses the Administration of not giving much more than rhetorical credit to the concept that the public has a serious and continuing interest in being informed. Whatever the rationale for classifying more and more information, such an attitude is clearly inconsistent with Congress's intention—to facilitate disclosure of information to those whom the government governs, and that the burden is on the government to justify closure, not on the individual who requests the material.[21]

Exemption 2 applies to matters "related solely to the internal personnel rules and practices of an agency." The first part is supposed to protect individual privacy and the second to protect law enforcement documents. In an important ruling, *Department of Air Force v. Rose* (1976), the Supreme Court tried to establish a public interest test.[22] Where there is a "genuine and significant public interest," disclosure is compelled *except* "where disclosure may risk circumvention of agency regulations." Student editors of the *New York University Law Review* had sought disclosure of Air Force Academy summaries of honors and ethics hearings for a law review article on disciplinary systems at military academies. The Air Force denied access on the basis of privacy, even though personal and other identifying information had been deleted from the requested documents. The district court granted the Department's motion for summary judgment, but the Second Circuit reversed and ordered disclosure if all pertinent identifying information could be deleted. The Supreme Court, in turn, found that the files did not contain the "vast amounts of personal data" that under Exemption 2 constitute a personnel file, thereby supporting Congress's intention—"to open agency action to the light of public scrutiny."[23]

Exemption 3 allows federal agencies to withhold records that are "specifically exempted from discovery by statute," provided that the statute clearly "requires that the matters be withheld . . . in such a manner as to leave the agency no discretion . . . or establishes particular criteria for withholding." On this, confusion has reigned, as courts have been confronted with the difficult job of weighing one federal law against another. In its 1976 amendments, Congress tried to narrow the scope of information that could be shielded. Courts, meanwhile, have permitted exemptions under the Consumer Product Safety Act, the National Security Act, various laws protecting CIA records, the Postal Reorganization Act, and, of course, the Privacy Act. But the courts have split over whether the Privacy Act is a statute under Exemption 3. "Given the hardening of the positions and, for the first time, a fully developed analysis on each side, the issue is ripe for resolution," according to the Duke reviewers. The Reporters' Committee has called this exemption the "catch-all" exemption and a major block to public access.

Exemption 4 protects "trade secrets and commercial or financial information obtained from a person and privileged or confidential." It has two prongs. If the requested material contains "trade secrets," the information is exempt from disclosure, and no further inquiry is necessary. But, if the documents contain only "commercial or financial information," their exempt status depends upon a showing of privilege or confidentiality. Trade secrets include secret formulae or customer lists, whereas commercial or financial information is confidential matter, disclosure of which "would be likely to cause substantial harm to the competitive position of the person from whom the information was obtained" or "impair the government's ability to obtain necessary information in the future."[24] The District of Columbia Circuit Court, to which many such cases are appealed, has adopted a restrictive definition of trade secrets, presumably to allow for disclosure of industrial safety data and other information of interest to the general public.

Some of the information sought, although not always found releasable, has been the gross income of three Boston housing projects, affirmative action plans and compliance reports, a report of an aircraft manufacturer's investigation of an air crash, government tests of hearing aids, documents detailing production costs incurred by defense contractors, data on television-related accidents gathered from manufacturers, a Department of Agriculture report on government discrimination in arranging housing loans, a film of fishing for tuna that results in the incidental killing of dolphins, a White House report on the government's program for the development of the supersonic transport, and documents pertaining to the merger of pharmaceutical companies. Such material, although much of it from the private sector, is of vital concern to the public, which needs a right of access nearly as broad as the right to government information.[25]

By law the Federal Trade Commission (FTC) is authorized to make public all information it has obtained with the exception of trade secrets and names of customers. The FTC is not prevented by the Trade Secrets Act from releasing confidential business information other than trade secrets and customer lists. Trade secrets, by FTC definition, must be of enduring and intrinsic value, primarily secret product formulae, processes, or other secret technical information. The court, in *Interco v. FTC* (1979), accepted that definition.[26] But, in response to the use of the FOIA by corporations themselves to garner information on competitors, a House-Senate Conference Committee amended the FTC Act in 1980 to exempt large areas of agency documents relating to pricing policies, product safety, and truth-in-advertising. In 1981, another conference committee exempted large areas of documents held by the Consumer Products Safety Commission, including safety and warranty data. Also in 1981, Congress amended the Omnibus Tax bill to exempt from disclosure the auditing standards and rules adopted by the Internal Revenue Service.

The continuing debate on Exemption 4 is naturally complicated by the private nature of American business and the public law constraints implicit in the Freedom of Information Act. In determining whether information requested contains trade secrets, the courts have traditionally applied the definition in the *Restatement of Torts*, which is overly broad:

> A trade secret may consist of any formula, pattern, device or *compilation of information* [emphasis added] which is used in one's business and which gives him an opportunity to obtain an advantage over competitors or suppliers who do not know or use it.

This can allow for a great deal of information to be exempted from public scrutiny, private property or not, and can further privatize material the public needs for the self-governing process.[27]

Exemption 4 has spawned a peculiar species of FOIA litigation—a "reverse" FOIA suit. Richard B. Kielbowicz, in a thorough study of the act and the government's corporate information files, explains that a person who submitted information to the government then seeks an injunction to prevent disclosure.

"Guidelines for reverse cases are unclear; decisions are sometimes inconsistent and confusing," Kielbowicz writes. Until the Supreme Court rules definitively or Congress acts decisively, he believes reverse FOIA suits pose one of the most serious obstacles to access to corporate information.[28]

Exemption 5, sometimes called the "executive privilege" exemption, prevents disclosure of "inter-agency or intra-agency memoranda or letters that would not be available by law to a party other than another agency in litigation with the agency." It has also been called the "nondiscoverable documents" exemption because, in granting FOIA access to information, Congress did not intend for the statute to become a way of circumventing civil discovery rules. That is, the exemption protects from disclosure documents traditionally accorded protection for certain evidentiary privileges in the discovery process. Such privileges include information acquired in an attorney-client relationship, and the exemption is meant to protect "predecisional communications," but not "communications made after the decision and designed to explain it."[29] The theory is that disclosure of memoranda generated before the deliberative process was complete might diminish the quality of decision-making. As Gillmor and Barron explain, "Advisers might be less candid if their recommendations were subject to public scrutiny."[30]

As an "executive privilege," Exemption 5 hides from public view working papers, studies, and reports that circulate among agency personnel before they reach a decision. Again, the purpose is to encourage frank debate. For example, the FOIA requires that university research grant applications and progress reports submitted to the federal government be made public on demand. Letters of evaluation, however, that are part of the peer review procedure, may be kept secret as part of intra-agency correspondence. In this, the exemption is meant to protect confidentiality and the deliberative process itself. Often the distinction between pre-decision and post-decision material is subtle and arbitrary. A Watergate Special Prosecution Force memorandum, as part of the group's required report to Congress recommending that President Nixon not be indicted, was ruled disclosable because it was part of the final opinion. By itself, however, the document would have been exempt from disclosure as a "pre-decisional intra-agency legal memorandum."[31]

A National Public Radio reporter was denied Department of Justice information on its investigation into the mysterious death of Karen Silkwood, a worker at a plutonium manufacturing plant. Using the FOIA, the journalist sought access to files marked "death investigation" and "contamination," the former denied on the basis of Exemption 5 and the latter on the basis of Exemption 7, which covers law enforcement records. Portions of the "death investigation" file were the working papers of attorneys, including notes and observations for personal use in analyzing evidence and legal issues. The exemption is not without merit, especially as it promotes the candid expression of ideas among high-level officials, but, as journalists in particular have found out, it can serve the privatization process as well.

Exemption 6 is the FOIA's explicit privacy clause, excluding from public view "personnel and medical files and similar files, the disclosure of which would constitute a clearly unwarranted invasion of personal privacy." Maxwell and Reinsch, in their review of cases falling under the privacy exemption, charge government agencies with trying to hide behind the exemption. In one case, the court, attempting to define "clearly unwarranted," said the exemption was designed to protect "intimate details of personal and family life."[32] In another, where the National Labor Relations Board denied legal scholars the names and addresses of workers eligible to vote in union elections, the court said that "clearly" was itself a "clear" instruction that it should "tilt the balance in favor of disclosure."[33] But, in a similar case, the court allowed that the Department of Health and Human Services could withhold a list of the names and addresses of its employees from the union that sought to communicate directly with them. While the court acknowledged that "collective bargaining is a matter of grave public concern," it said that any benefit from disclosure would "inure . . . to the union, in a proprietary sense, rather than to the public at large." Because disclosure of the information might subject the workers to mailings and personal solicitations, the court believed that there was a "strong privacy interest" in keeping the home addresses confidential.[34] In a third case, the court upheld an order compelling the Environmental Protection Agency to reveal documents that included the names and addresses of homeowners whose building lots contained uranium deposits.[35]

Chief Judge C. J. Winter, of the Fourth Circuit Court of Appeals, vigorously dissented in the public employees case, arguing that the "right to privacy in one's home address is an interest of little value." He said the disclosure was appropriate because Congress has recognized that labor organizations and collective bargaining in the civil service are in the public interest. He went on to note that because communication with those employees that the union is obligated to represent is essential to effective representation and that disclosure of the lists would serve both the union's and the public's interests. Rarely is personal privacy invaded by such mundane information, details of employment and abode that the average person does not usually care to keep secret.[36]

Since the FOIA was amended in 1974 to require *in camera* inspection, the results at least have been more thoughtful and judged on the merits of each situation. But the obligation still favors the keeper of government information, because the requestor has no way of ever learning what information remains exempt. As one court noted the problem, one party knows the contents of the withheld records, while the other does not, and the judiciary is charged with the responsibility of deciding the dispute without altering that unequal condition, because that would involve disclosing the very material sought to be kept secret.

Not only does the *in camera* solution fail to alleviate the basic problem of public access (although it tries), it also limits the individual's control over the disclosure of her or his records. Although the court's review may be less protective of government secrecy and weigh more in favor of public interest, individuals are not permitted to have a say in their personal privacy under the FOIA. As Maxwell and Reinsch rightly assert, the exemption was created to strike a balance between the individual's right to privacy and the public's right to know, but it failed to take into consideration the possibility that individuals would not object to the release of information about them. Or, if they do object, it may not be against the release of all information. Besides, if personal privacy were of genuine government concern, that interest can best be expressed by involving the individual in the decision to release such information. "Agency personnel, as custodians of the records, assume the responsibility for those decisions," write Maxwell and Reinsch. "They can and do deny access on behalf of the individual, providing simultaneously an avenue for the agency to protect its own privacy." Who is better qualified to decide when an invasion of privacy is "clearly unwarranted" than the person whose privacy is being invaded?[37]

Exemption 7, which shields "investigatory records compiled for law enforcement purposes," is supposed to control the flow of information that, if known publicly, might seriously harm ongoing proceedings of a civil, criminal, administrative, or judicial nature. When first applied, the exemption protected a wider than intended amount of information from the public. But, with the amended 1974 version, the government is required to show that the requested documents are truly investigatory and compiled specifically for law enforcement purposes. The government must also prove that disclosure would result in one of six "categories of harm":

1. That disclosure would interfere with law enforcement proceedings;
2. That the information would deprive a person of her or his right to a fair trial or an impartial adjudication;
3. That publicity would constitute an unwarranted invasion of privacy;

4. That disclosure would reveal the identity of a confidential source and confidential information;
5. That the information would disclose investigative techniques and procedures; or,
6. It would endanger the life or physical safety of law enforcement personnel.

All FBI investigatory records are, in the language of Exemption 7, "compiled for law enforcement purposes," the breadth of which recently divided the Supreme Court in an access case. In *FBI v. Abramson* (1982), the court upheld FBI claims under the exemption against a journalist's requests for the Bureau's documents on certain Nixon administration critics.[38] Among those on the White House "enemies list" were John Kenneth Galbraith, Reinhold Niebuhr, Cesar Chavez, and Dr. Benjamin Spock, whose names appeared in summaries compiled from FBI files in 1969 at the request of the White House. Howard Abramson requested the summaries in 1976 and argued that once the information was taken from the FBI files and sent to the White House in summary form it was no longer an investigatory record compiled for law enforcement purposes. But the Supreme Court disagreed in a five-to-four decision.

In another narrowing of the FOIA through this exemption, the Supreme Court held that information initially contained in a record made for law enforcement reasons continues to meet "the threshold requirements of exemption even where that recorded information is reproduced or summarized in a new document prepared for a non-law enforcement purpose." Associate Justice Brennan, in a sharply worded dissent, accused the majority of rewriting the statute, and he stated that the Court had rejected a logical and straightforward interpretation of the law and had replaced it with one less plausible.

Associate Justice O'Connor charged the majority with rewriting Exemption 7 to conform to its concept of public policy. She said that the exemption's legislative history left the Court "no reason for overriding the usual presumption that the plain language of a statute controls its construction." With her three dissenting colleagues, Justice O'Connor agreed with the District of Columbia Court of Appeals, which had reversed a lower ruling, that the documents had been compiled for political, and not law enforcement, purposes. Justice White, however, thought O'Connor's position "rhetorical and beside the point."

But the same may be said of the majority's position, which assumed that such information had been collected to enforce some mysterious law, and implied that some information may be termed classified at birth. As was noted in the other FOIA exemptions, the original purpose for which information is compiled by the government should be the major factor in determining public access. The Court in *Abramson* in effect privatized all FBI files.

It is also interesting and disturbing that, in a footnote, the court expressed concern over the flow of essential information to the government being slowed by the Court of Appeals ruling. Yet these are precisely the reasons cited time and again by thoughtful legislators for enacting the Freedom of Information Act—to end bureaucratic secrecy and to guarantee public access to government information. The explicit privacy exemptions, Exemptions 6 and 7, were meant to protect the personal privacy of individual citizens, not the secrecy of the government agencies that serve them.

Exemption 8 protects information "contained in or related to examination, operating, or condition reports prepared by, on behalf of, or for the use of an agency responsible for the regulation or supervision of financial institutions." Similar to Exemption 4, it prevents disclosure of financial reports or audits that might jeopardize public confidence in banks, trust companies, investment firms, and other financial institutions. A New York district court observed, for instance, that correspondence between a bank and the Federal Reserve Board would probably fall under the exemption. But a court also held that a Securities and Exchange Commission study of a broker-dealer trading problem did not fall within Exemption 8. House and Senate reports on the FOIA indicate only that Congress wished to avoid the harm that disclosure of sensitive financial information might bring.[39]

Exemption 9 excludes "geological and geophysical information and data, including maps, concerning wells" in an effort to prevent speculation based on information about the location of oil and gas wells. The Federal Power Commission used the exemption to deny Ralph Nader access to FPC and American Gas Association estimates of natural gas reserves. Nader contended that Exemption 9 only applied to data and maps that could benefit a competitor. The FPC, in response, said that estimates of reserves were based on such data, and indeed could be useful to competing firms.[40] In other situations, the FPC ordered the raw data for computing estimates of companies' natural gas reserves to be entered on the public record. When the Pennzoil Company sought judicial review of one such order, the commission conceded that disclosure might benefit competitors, but added that "full disclosure of the facts upon which the rates are to be determined" was in the public interest and thus outweighed any damage to the company. On remand, the court asked the commission to delete any possible identifying details.[41]

What continues to haunt a democratic society, especially one so historically committed to both privacy and publicity, is the enormous potential for masking either commitment under the guise of the other—that is, privacy as the ruse for limiting or avoiding publicity. Even in those cases that reached litigation, and where the courts opted for public disclosure, there remained a strong desire for secrecy, for private concerns, for what's good for the individual rather than what's good for society in general. These FOIA situations also suggest that legislated privacy or publicity, especially in cases involving closure or disclosure of public information, is the preferred way to deal with the access question. This is because the judicial system is designed to *interpret* the law, rather than to *follow* or *enforce* that which legislators have prescribed. Adjudication has always been necessary, however, even in revolutionary societies but most assuredly in the American system, which West Virginia Chief Justice Richard Neely characterizes as the "measured straining in opposite directions."[42] And perhaps there is no better way, but the opinion of this writer is that the strain should be more in the direction of, to paraphrase Woodrow Wilson, open covenants arrived at openly.

After charting the rapid growth of the "information society," as the world of the 1980s has been described, Congress in its wisdom decided to attempt to control the flow of information. Some legislators have advanced a more liberal view of needs, whereas others have advocated a more conservative approach, but all seem to be saying that the collector of information, not the requestor or even the person on whom the information is compiled, is the one whose privacy is most likely protected. The fact that the FOIA has no provision for individual determination of unwarranted invasion is at least some proof of the pudding. Legislation designed to open government at the same time it protects privacy has in fact accomplished just the opposite—the FOIA ensures neither openness nor privacy. When requestors, such as persistent journalists who know how to use the FOIA, have forced compliance, they can testify, with William Safire, that the Freedom of Information Act "has done more to inhibit the abuse of Government power and to protect the citizen from unlawful snooping and arrogant harassment than any legislation in our lifetime."[43]

When the House Government Operations Subcommittee examined how the government's major departments were obeying the FOIA, it found that, excepting procedural denials, 91.9 percent of the requests processed by the agencies were granted in full in 1984. Representative Glenn English, who headed the subcommittee, told the *New York Times* correspondent David Burnham that the FOIA is "successfully accomplishing its primary purpose: making government documents available to those who want them." But the chairman cautioned against accepting the numbers as a comprehensive assessment of compliance with the act, for, as he reported to Burnham, the statistics do not show whether the agencies are processing requests in a timely fashion (which is important to the news media), whether the agencies are granting fee waivers to those entitled to them, or whether all disclosable information is actually disclosed. Nor did the subcommittee have figures on the number of contested cases in litigation.

The subcommittee found great variation in the response of the eight cabinet-level departments it examined. As reported by Burnham, the two with the best records were the Department of Health and Human Services, which granted 98.9 percent of the requests it processed, and the Department of Defense, which granted 92.1 percent. The agency with the worst record was the State Department, where only 29.1 percent of the requests were approved. English said that there still was considerable room for improvement in the handling of information under the law.[44]

10

Data Privacy

"The Privacy Act, if enforced, would be a pretty good thing. But the government doesn't like it. Government has an insatiable appetite for power, and it will not stop usurping power unless it is restrained by laws they cannot repeal or nullify." These words came from former Senator Samuel J. Ervin, Jr., during an interview with David Burnham for Burnham's thought-provoking book, *The Rise of the Computer State* (1984).[1] His is indeed, as the subtitle promises, "a chilling account of the computer's threat to society," and it provides an appropriate backdrop of events leading up to Congress's enactment of the Privacy Act of 1974. The book is also a stirring sequel to Arthur R. Miller's earlier *The Assault on Privacy* (1971), which is said to have sparked the Senate's Constitutional Rights Subcommittee to convene congressional hearings early in the spring of the year it was published. Chaired by the late Senator Ervin, the lengthy set of hearings focused on federal databanks, computers, and the Bill of Rights. The result, three years later, was the Privacy Act.

Building his case for personal privacy from within the larger frame of the recently arrived information age, Burnham views the loss of privacy as symptomatic of an even more fundamental social problem: the growing power of large public and private institutions in relation to the individual citizen. He examines in

great detail how the widely acknowledged and heavily adver-
tised ability of the computer to collect, organize, and distribute
information tends to enhance the power of the bureaucratic
structures that harness the computer to achieve their separate
and often worthy goals. But Burnham's study is not just about
computers, nor is he opposed to their use; it explores the ways
the computer, as utilized in society, affects our values and influences the way we think about what is important. Burnham pon-
ders the issues, and his findings are relevant to the present
treatise, which has considered personal privacy in relationship
to public openness. In this chapter we continue to examine the
effects of information on privacy.

Paul Sieghart, a British writer on the subject, describes how
the advent of the electronic computer, a rather unintrusive "assembly of elegant boxes sitting peacefully in someone else's
building," has brought about the recent revival of concern over
privacy. The field on which the debate about privacy and computers is conducted is not the field of intrusion or of surveillance
as such, Sieghart points out, but the field of privacy of *information*.[2] Thus, Alan Westin's definition has become the accepted
one: "the claim of individuals, groups or institutions to determine for themselves when, how, and to what extent information about them is communicated to others."[3] Or, as Professor
Miller abbreviates the definition, privacy is "the individual's
ability to control the circulation of information relating to him."[4]
The term common to all definitions these days is "information."
Privacy as solitude and seclusion, or the right to be let alone, or
to be free of surveillance and intrusion—all traditional concepts—have given way in large measure to the fear of informational invasion of privacy. And that's where the computer, now
seen as a not-so-elegant threat, has become central to the
debate.

"Computers," Burnham summarizes, "have allowed for
more organizations to have far more access to far more people at
far less cost than ever was possible in the age of the manual file
and the wizened file clerk."[5] Like the telephone, the telegraph,
and the printing press before it, the computer has been an in-

strument of fundamental societal change. A growing number of corporations, government agencies, trade associations, and a myriad of other institutions now have access to detailed computerized data on most citizens. One simple catalogue purchase by the typical American household quickly results in a dramatic increase in the amount of computer-generated mail.

What the Postal Service delivers daily to homes and offices is but the tip of the information iceberg. The computer enables, as well as encourages, ways of reaching and keeping tabs on the nearly 250 million people of the United States. Worldwide, the barrage is proportionately the same. Miller puts it this way: "In a computerized environment, anyone who knows a credit grantor's identifying code number and has access to a telephone may be able to reach the reservoir of detailed financial information that already exists on over a hundred million persons.[6] That figure has increased substantially since Miller made the observation in 1971.

Government is not necessarily the worst offender, but it is the single biggest collector and distributor of information about citizens. This itself increases the probability that such data may be acquired and used under questionable, if not illegal, circumstances. History is filled with instances of government taking liberties with its surveillance capability. Because bureaucracies by definition are powerful and seek to enhance their hold at every opportunity, computer technology makes it easier for our worst totalitarian tendencies to go undetected.

Burnham relates that it was not until some years after the fact that the American people learned of the Kennedy administration's surveillance of certain civil-rights activists, or of the Johnson administration's secret tracking of antiwar protestors, or of the Nixon administration's use of computerized tax records on individuals the President did not like. Today, when people congregate to complain about their loss of personal autonomy, they focus on data intrusion by the government and other organizations with large centralized and interlocked information systems. "How did Macy's find my unlisted number?"

The computer, although not always the handy scapegoat for

human error, may be compared to the gun that is available to a variety of potential users. The difference in use will depend upon the intent of the user. As Burnham so nicely illustrates, automobile manufacturers opposed safety legislation by arguing that poor drivers, not poorly designed cars, were the cause of highway accidents. This view has changed, but the National Rifle Association still adheres to the slogan, "Guns don't kill people, *people* kill people." So it is with the computer, a harmless productive invention in the hands of users whose intentions are beneficial to society. In the hands of others less honorable, the computer can become an instrument of destruction.

Miller raises another concern: that widespread computerization of personal data coupled with continuous demands for data by information managers will "slowly narrow the community's conception of what is private."[7] Progress is frequently accompanied by a tendency to alter, and in some cases to erode, the history and significance of certain basic values. Despite occasional concern over highway slaughter, for example, the public has become accustomed to this new way of life. The 55-m.p.h. speed limit would seem to indicate public interest in reducing the number of accidents. A more direct solution, however, would be reducing the power of the combustion engine and not the speed limit. But the motoring public, which cherishes its automotive freedom (and privacy), is not likely to press for that simple solution. With some alarm, Miller is concerned that people accustomed to the revelation of sensitive personal data may eventually define most information as public and place it beyond the law's protection. By way of analogy, he cites the narrowing scope of what is "obscene" as an instance of the community becoming accustomed to more explicit presentation of sexual themes.[8]

The computer, the gun, the car—all marvelous inventions—have been at the center of changes in human behavior, not all of which have been beneficial. "Modern technology seems to enable man to gain control over everything but technology," said Professor David F. Linowes, who was chairman of the U.S.

Privacy Protection Commission established by the Privacy Act.[9] In the context of the changing historical concepts of American individualism, as presented in Chapter 7, one cannot help feeling apprehensive about how the computer will affect the individual and her or his sense of community.

Already the term "privacy" is being used as a catch-all to denote, in addition to the older concepts discussed earlier, the sense of alienation or estrangement of the individual from society, especially government. Personal privacy begins to appear as alone-ness and loss of self, which can lead to defensive posturing, increased self-protection, self-indulgence, and self-interest. As individuals experience either too much information or a threat of the invasion of their private data, they reject the information and turn inward to shield themselves against intruders. Disappointment, says the economist Albert O. Hirschman, is a major factor in individual and societal preference changes. Frustrations over participation in public life inevitably redirect our energies toward the private life, and vice versa. But, again, it is not the strictly private existence that enhances and enriches the quality of life in general. Jacob Burckhardt's observation of 1943 retains the ring of truth—that after a long immersion in purely private concerns, the discovery of action directed to a public purpose constitutes a liberating experience, "a way of rising above the self-seeking of the individual and the family." Hirschman suggests that the instability associated with public action has a viable counterpart: "The greatest asset of public action is its ability to satisfy vaguely felt needs for higher purpose and meaning in the lives of men and women."[10]

As we change or learn to adjust, we also need to create more accurate terms to express the various notions often addressed under the privacy rubric if they are to be properly studied and public policy developed. For example, in Chapter 7 it was suggested that the new right of publicity/personality more properly belongs, in theory and in practice, within the law of property because privacy is not in question. An intellectual property approach will help avoid potential First Amendment conflicts by limiting celebrities' lawsuits to those situations where their business rather than personal interests are at stake. Other notions of privacy also beg for rethinking—at least to the extent that legal protection is deemed necessary. Thus, Professor Miller's concern over society's hasty adjustment to more limited versions of privacy may be turned into a golden opportunity not only to revise our thinking on privacy but to revive our commitment to public matters.

Generally speaking, the concept of privacy, as old as human history, tries to distinguish between the individual and the collective, between self and society. The concept is based upon respect for the individual, and has evolved into respect for individualism and individuality. To maintain the continuity of that moral tradition, there must be some restriction on conduct from the outside, from other individuals as well as from government, that would destroy identity, individuality, or autonomy. In response to that position, legal protection of personal privacy has progressed on at least two levels: 1) law dealing with interference by government with the citizen's right to privacy; and 2) law directed not at government but against invasion by private individuals, groups, and organizations.

On the first level, protection comes largely from the Constitution and federal statutes like the Privacy Act. Some states, too, have enacted public-sector privacy laws. Minnesota led the way with the first omnibus bill, passed in April 1974, pre-dating by several months the federal law. By 1980, eight other states had enacted some form of privacy legislation designed to regulate government data flow, and state commissions on privacy have been established in Minnesota, Indiana, Illinois, and New Jersey. But the fact that most states have not enacted specific legislation does not mean that they place less value on privacy or that their citizens are less protected from government intrusion. In New Jersey, for example, where lawmakers created a privacy commission in September 1977 but no privacy laws *per se*, it was recently learned that the federal Privacy Act covers citizens when their privacy is in question at the state level. A

federal district judge ruled in September 1985 that it was unconstitutional for the state to require that more than 100,000 residents give a water company the names of every individual living
in their homes as part of the state's rationing plan to ease its
water-supply emergency. Judge Clarkson Fisher said that because the information could provide clues about "familial relationships and other intimate living arrangements," the freedom
from compelled disclosure was within the right of privacy as
recognized by the courts. The judge further ruled that the request that New Jersey residents receiving their water from the
company provide it with their Social Security numbers violated
the federal Privacy Act.[11]

The First Amendment is the traditional guarantee of private
religious freedom, as well as speech and press freedom. Also,
the Amendment's "right of the people peaceably to assemble,"
the right of association, is peripherally related to data privacy.
The Third protects the intimacy of private dwellings. The Fourth
limits searches, seizures, and arrests, and the Fifth alludes to
privacy in terms of self-incrimination and due process. One may
aver that the Constitution and its amendments, and especially
the Fourteenth, form a composite statement about the privacy of
the individual. The people have the right, as stated in the Fourth
Amendment, "to be secure in their persons, houses, papers,
and effects." The Constitution thus protects one's body from
assault, private places from trespass, and personal property
from theft.

On the second level of legal protection, common law and
legislation provide varying degrees of protection to individuals
against commercial and noncommercial use of their names, like-
nesses, and personal information. Because the government may
not be involved directly, the Constitution is not the direct source
of these rights. We have laws affecting trespass, unsolicited
mailing of sexual material, street demonstrations, and conduct
that constitutes "action" rather than "expression." Just as the
Constitution has been called a document of principle, it may be
said that legislated law is a matter of policy. Thus, data, in their

various forms, are at once matters of constitutional principle *and* matters of legislated policy. But, as Ronald Dworkin, the legal theorist who developed the differing concepts of law, insists, principle must prevail when the two arguments are in conflict. Data privacy, as it results in ephemeral legislation meant to correct specific abuses of information, may have to give way to a more utilitarian constitutional principle whose purpose is less transitory.

A series of related questions come to mind. Can the government constitutionally or legislatively collect, maintain, use, and disclose the contents of official databanks and files about individual citizens? What laws, regulations, or rules govern the disposition of official arrest and health records? What personal data may legally be required to be disclosed as a condition of government employment or benefits? Are public employees less protected in their personal privacy than are private workers? And, because the Constitution does not cover private databanks or other nongovernment data-gathering, what are the individual's rights regarding common law or legislation precedents? History suggests some answers.

For centuries physical surveillance was the only real threat to privacy, except, of course, the dossiers monarchies kept to control population movements and admonish disloyal subjects. Churches, too, used this method to keep their flocks thinking and behaving as one. (History at times appears to repeat its less than glorious episodes, as when President Nixon, in the tradition of the kings, used official surveillance information to punish disaffected citizens!) Physical invasion of privacy was possible by actual entry onto property, eavesdropping by ear, and the overseeing of individuals and groups.

In America, the Fourth Amendment was adopted to control that kind of surveillance by stipulating that searches and seizures must be "reasonable" and decided by judicial process. To protect citizens against such methods of psychological surveillance as torture and self-incrimination, the framers designed the Fifth Amendment to prohibit individuals from being compelled to testify against themselves and to enjoy the due process of law.

Until the late nineteenth century this constitutional framework for protecting privacy seemed adequate. The reasonable search and seizure provision permitted an acceptable balance between individual and group claims to privacy and the government's need for information. But new technology—first the telegraph, then the telephone—changed that balance by extending personal conversation beyond the home and office and eroded physical boundaries around privacy. With the appearance of the polygraph and the widespread use of personality tests, psychological boundaries also began to fade. Although neither is admissible in court, the tests were, and are, used for job selection and other nonjudicial decisions.

Data surveillance, as an acceptable procedure in government as well as in private-sector affairs, gained momentum between the world wars, when record-keeping evolved as the natural outcome of an industrial society with a growing regulatory and welfare function by government and a growing bureaucracy in private organization life. The immediate legal protection that followed was the setting of general standards of confidentiality for information given to public agencies under compulsion of law. But, as data surveillance far outdistanced either physical or psychological surveillance, such legal protection soon became inadequate.

Computers, as we have seen, made possible the collection, storage, manipulation, and use of vast amounts of data, much of it arbitrary, undiscerning, and indiscriminate. While technology made it possible to collect information at an ever increasing rate, an even greater effort was made to centralize data systems that could be shared by all segments of society. The concentration of diverse but related information ("related" to the person or group) may affect community attitudes toward privacy, for, among Professor Miller's concerns is the possibility that the public may view the composite dossier differently from the way it may see the same information before it was brought together.

If confidential medical data were mixed with less sensitive information, there might be the tendency to treat it as data of low sensitivity.[12]

What all this means in terms of legal protection for individual and group privacy is not simple. First, there is the obvious need, as noted earlier, to redefine what we mean by privacy. Societal notions, or senses, of privacy have always undergone change. Second, there may have to be new laws, regulations, and rules that apply to computer-driven devices that have the capacity to invade data privacy. The more information pools are centralized, the more serious the unrestricted flow of information becomes. American society wants more and better information *and* personal privacy; as usual, we want it both ways.

Privacy, as we have seen its history evolve, has come to mean nearly the opposite of what it did, let us say, three hundred years ago. Hirschman, the economist, points out that during the Renaissance the stress was on civic virtue and involvement in public affairs. Today, society stresses the pursuit of private self-interest as most conducive to a well-regulated social order. As recently as the American eighteenth century, a term such as "happiness," an almost wholly private venture today, still had a substantial public dimension. When Thomas Jefferson, in the Declaration of Independence, designated the "pursuit of happiness" as an inalienable right, he had in mind the *public* happiness, that is, a performance of economy and society that is satisfactory to its members, indeed, the test and justification of any government.[13]

This tracking of the concepts—public and private—helps explain the current popularity of privacy and, thus, the great concern over its erosion. Conversely, were today's society more public and public-serving, the individual would be less inclined to put a premium on her or his privacy. Whether the once-dominant public realm ever regains its position in national life is a matter for the future, not history, but the philosopher William M. Sullivan interprets with guarded optimism the current American retreat to privatism as a continuing search for what counts in life, "a hunger for orientation that neither the dynamics of capitalist growth nor the liberal vision of politics provides." Sullivan, in his imaginative *Reconstructing Public Philosophy* (1982), writes:

> Those who read the American spirit as so dominantly individualistic that private comfort and competitive achievement define a monochrome of national traits are both dangerously distorting our past and threatening our future, because they are closing off a sense of that living civic tradition which has been and continues to be vital to national life.[14]

In addition to a historical yearning for privacy, there has been as strong a craving for, in Sullivan's words, a "life of inclusion in a community of mutual concern." Large-scale social processes cannot remain merely technical issues, but must be understood as part of the texture of private living, just as private life is woven into the fabric of public organization. The invasion of privacy by computer systems is no mere mirage, but it is also one of those technical issues that can retard genuine social progress.

As a prelude to the 1976 Bicentennial, governors of the thirteen original states met in Philadelphia on September 5, 1974, to commemorate the 200th anniversary of the convening of the First Continental Congress. During the two-day conference, the governors drafted four resolutions—two of them routine, the others controversial. They passed a declaration in commemoration of the first Congress and a resolution on the bicentennial itself. The state leaders then passed resolutions that dealt with basic problems inherent in contemporary America. The Resolution on the Quality of Life declared that all citizens enjoy the basic right to a decent job, "commensurate with the promise of the Constitution."

The Resolution on the Right to Privacy called for the passage of a constitutional amendment to include proper restraints on all public and private information-gathering agencies and on the dissemination of criminal justice information in order to further protect the individual citizen's privacy. Discussion was heated and turned the conference into what one observer called "a virtual re-creation of the debates 200 years ago."[15]

The genesis of the privacy resolution was in the White House years of Presidents Johnson and Nixon, when various federal agencies conducted computer surveillance of civil rights activists and those citizens opposed to the Vietnam War. Especially alarming was the FBI's National Crime Information System, a massive databank of information accessible through computer terminals in police stations across the country. Massachusetts and Pennsylvania refused to plug into the FBI's criminal history program until adequate safeguards were provided to protect the states' citizens. Both also refused to cooperate with the Drug Abuse Warning Network (DAWN), a computer tracking system designed to establish instantaneously the whereabouts of known or suspected drug users. As Massachusetts Governor Francis Sargent told his colleagues at the Philadelphia conference: "We can no longer complacently stand by as federal officials investigate our tax returns, monitor our telephones, review our banking history, survey our medical files—not only without knowledge but for no legitimate public purpose."[16]

During the debate in Philadelphia, the governor of New York said that the privacy resolution "overstates the right to privacy," and another leader felt that the wording contained a potential abridgment of press freedom. Others thought that the right to privacy was implicit in the federal Constitution, as various Supreme Court rulings had begun to show. After a day and a half of debate, the conference approved a resolution calling on Congress "to take action with respect to the potential threat to personal freedom and the right to privacy by uses and misuses in the collection and dissemination of data concerning or relating to private citizens."[17]

Congress reacted almost immediately. Out of concern for states' rights and also the growing public interest in data privacy, Congress adopted a number of measures to cope with the sophisticated information-gathering procedures and mechanisms. Title III of the Omnibus Crime Control and Safe Streets

Act of 1968, as a beginning, limited private wiretapping, eavesdropping, bugging, and other willful interceptions of oral and wire communications. The Family Educational Rights and Privacy Act (FERPA) of 1974, also known as the Buckley amendment (for Senator James Buckley of New York) to the Elementary and Secondary Education Act, gave parents the right to inspect school records of their children, including intelligence and psychological test scores, health information, family background data, teacher and counsellor ratings, and observations and reports of serious or recurrent behavior patterns. The act also enables parents to challenge and correct school records. But the three major pieces of federal legislation designed to protect informational privacy were the Fair Credit Reporting Act of 1970 (FCRA), the Privacy Act of 1974 (PA), and the Right to Financial Privacy Act of 1978 (RFPA).

The FCRA, which went into effect in April 1971, is meant to correct abuses and misuses of consumer credit information, and is considered the federal government's most significant effort to regulate data collected and transferred between private parties. The FCRA requires that credit reporting agencies adopt reasonable procedures in providing credit information in a fair and equitable manner and with deference to "confidentiality, accuracy, relevancy and proper utilization of the information."

Under the statute, a consumer reporting agency that also collects data on employment and insurance must disclose to the consumer the "nature and substance of all information (except medical information) in its file on the consumer" upon request. If the consumer finds the data inaccurate or incomplete, the agency must reinvestigate and record the correct status of the information or face a civil action for "willful noncompliance." Thus, it is the individual, not the business for whom the data are compiled, that is protected, thereby emphasizing "respect for the consumer's right to privacy."

Because the obtaining of credit has become a major event in every American's life, one would think that the FCRA would ensure accuracy as well as privacy. It is the precision of credit information, not personal privacy, that is the issue, although in a consumer society the two are linked. But the FCRA has been criticized on a number of counts. First, it is vague on who is allowed access to the data, as it permits disclosure to "anyone with a legitimate business need for the information in connection with a business transaction involving the consumer." Although the consumer has the right of access to check for accuracy, and therefore appears to control the data, she or he has no say about to whom the information is disseminated.

Second, the consumer does not have the right to inspect the credit file *in person*. The FCRA only requires the agency to provide the individual with "the nature and substance of all information," a summary, in other words. Nor does the act permit the consumer to correct information while it is being acquired. The summary is available only after all the information has been filed in the computer.

Third, critics point to a major loophole: that the act is directed only at credit reporting agencies, and not other organizations that may acquire a person's financial record and so not be bound by the FCRA. Again, it is not so much the private nature of the data that worries most consumers, because to secure a bank loan, an insurance policy, or new employment, the average consumer is more than eager to reveal such personal information. Rather, it is the accuracy of the information that most concerns a person in these circumstances, and the FCRA does not go far enough in that direction.

After enacting the Privacy Act in 1974, Congress returned to the issue of financial privacy with the Right to Financial Privacy Act in 1978, an effort to correct an imbalance in privacy created by the Bank Secrecy Act (BSA) of 1970. This earlier law—anything but a secrecy or privacy law—required banks to keep records, and to file certain reports that have "a high degree of usefulness in criminal, tax, or regular investigations or proceedings." Together with certain Treasury Department regulations, the BSA required financial institutions to keep records of the identities of customers and to store microfilmed checks for five years. The law also required reports on other documents and foreign currency transactions.

In *California Bankers Association v. Shultz* (1974),[18] the plaintiffs, which included several bank customers and the American Civil Liberties Union, unsuccessfully challenged the constitutionality of the law. They contended that the BSA and its implementing regulations violated the Fourth Amendment's protection against unreasonable search and seizure. They said that the act also violated the First, Fifth, Ninth, Tenth and Fourteenth Amendments, and the banks added the extra work involved and said the law made them agents of the government in the surveillance of citizens. Associate Justice Douglas, vigorously dissenting from the Supreme Court majority, alluded to this concern: "It would be highly useful to governmental espionage to have like reports from all our bookstores, all our hardware and retail stores, all our drug stores. These records too might be 'useful' in criminal investigations."

Two years later, the Court was more explicit in its thinking, when, in *United States v. Miller*,[19] it ruled that the customer's bank records were not owned by the customer but by the bank. The Court said that it perceived no legitimate "expectation" of privacy in a person's checks or deposit slips or in microfilm of the same items. "The checks are not confidential communications, but negotiable instruments to be used in commercial transactions." Furthermore, the Court noted that bank statements, although they may contain personal information, "contain only information voluntarily conveyed to the banks and exposed to their employees in the ordinary course of business." When a depositor uses a bank, the Court concluded, she or he "takes the risk, in revealing . . . affairs to another, that the information will be conveyed by that person to the government." Again, however, as with all legislation meant to protect data privacy, the prior issue would appear to lie with the "privacy" at stake, the "intimate" or "personal" information that one expects to be secured from public disclosure.

The *Miller* decision, although it may appear to be out of step with the current societal shift away from public welfare, is at least a principled ruling, to borrow from Ronald Dworkin's theory. The justices seem to have reached their position without distinguishing among economic classes and without apparent concern for the political ramifications of their decision. That is, rich and poor alike are subject to the openness of financial transactions. Ironically, the poor may enjoy more economic privacy because they may not be able to afford the luxury of banking. *Miller* is also in keeping with the Supreme Court's reluctance over the years to countenance constitutional protection for what amounts to commercial speech, as opposed to the more important political speech.

Dworkin, whose philosophy on legal rights was presented in Chapter 5, might observe that, when legislators opt to constrain government access to personal information, they do so because they believe the community as a whole would be better served by the constraints. That is Dworkin's basic policy argument (as opposed to his argument for principle), and legislators, not the judiciary, are better equipped as representatives of the populace to make public policy. With *Miller*, the Supreme Court in effect upheld the Bank Secrecy Act, which was designed, according to the author of the Senate's version, William Proxmire, "to provide law enforcement authorities with greater evidence of financial transactions in order to reduce the incidence of white-collar crime."[20]

Dworkin's argument of principle is based on rights; arguments of policy are based on goals. A "policy" is a standard that sets out a goal to be reached, "generally an improvement in some economic, political, or social feature of the community." A "principle" is a standard to be observed, not because it will advance or secure an economic, political, or social situation, but because "it is a requirement of justice or fairness or some other dimension of morality." For example, the standard that automobile accidents are to be reduced is a policy, and the standard that no man may profit from his own wrong a principle. Whereas a principle may state a desired social goal, a policy may state a principle, or "by adopting the utilitarian thesis that principles of justice are disguised statements of goals (securing the greatest happiness of the greatest number)."[21] Ideally, the courts determine rights and legislatures decide goals; hence, the government's role in privacy is often double-binding, for it is at once responsible for private rights *and* public goals. To attach Dworkin's theory to financial privacy is not to diminish the complexity of rights; rather it is to point out once more the conflicting nature of personal privacy in a public society. *Miller*, in upholding the government's effort to reduce white-collar crime, established a policy standard to be observed; in separating one's personal privacy from one's personal pocketbook, it also set a good moral principle.

We come now—finally—to *the* Privacy Act, enacted in 1974 during the same session that Congress amended the Freedom of Information Act. In both, Congress was responding to the growing threat to privacy brought on by the increasing sophistication of computers. Section 2 of the act states that the increasing use of computers and sophisticated information technology, although essential to the efficient operation of the government, has greatly magnified the harm to individual privacy. And it states that the right of privacy is a personal and fundamental right protected by the Constitution. The act then says that it is necessary and proper for Congress to regulate information systems maintained by federal agencies.

The Privacy Act, as the federal counterpart of the FOIA, placed certain information constraints on government agencies. In its final form, however, it did not go as far as Senator Ervin would have liked. As originally proposed, the act was to have covered the information systems of private industry and of state and local governments, in addition to those at the federal level. As approved by Congress, the law only applied to federal agencies and the private organizations with which they did business. The senator's plan also called for the act to include most of the government's intelligence and law-enforcement activities, but the final bill eliminated those agencies from many of the requirements of the Privacy Act.

Instead of a permanent independent commission to monitor and regulate the development of significant new computer systems, Congress authorized the creation of a temporary privacy protection commission to study the various data systems, to recommend standards and procedures, and then go out of business. Interestingly, when the commission issued its report in July 1977, which included numerous recommendations, it mainly urged self-control because of what its chairman Professor Linowes of the University of Illinois described as existing over-regulation and overlegislation in the area of privacy.

What the act protects generally is the government's disclosure of individual records without the individual's prior written consent. But, because it was believed that such an absolute rule would unduly hinder the functioning of government, eleven exceptions were written into the Privacy Act. The first exception allows information to be disclosed to "officers and employees of the agency which maintains the record who have a need for the record in the performance of their duties." Consent is not required for such intra-office use.

The second exception allows for disclosure as required by the FOIA, or whenever an individual wants to correct her or his record. The third permits disclosure for "routine use" without consent, but such use is supposed to be compatible with the purpose for which the data were initially collected. Other exceptions deal with information gathered for the Bureau of Census, for statistical research, for the National Archives, for law enforcement, for health and safety, for either house of Congress, for the Controller General, and pursuant to a court order. The act requires that an accounting be made for each disclosure, except inner-agency use and release necessitated by the FOIA.

Unlike exemptions from the FOIA, exceptions to the Privacy Act are not automatic. Every exception must be published in the *Federal Register*, and no system of records can be excepted from all provisions of the act. Each government agency is required to publish those systems for which an exception is claimed, the specific part of the statute from which the records would be excepted, and the basis for claiming the exception. One provision of the act requires that federal agencies collect and maintain only information that is "relevant and necessary" in order to meet their legal obligations or purposes stipulated by the White House.

Another provision requires that, to the maximum extent feasible, federal agencies obtain their information directly from the individual in whom they are interested. A third provision requires agencies to inform the individual of the legal authority under which the data are being sought (whether compliance is voluntary or mandatory) and how the information will be used. A fourth provision gives every individual the right to see and correct her or his records. Regarding this right of inspection, the Privacy Act did not set up a mechanism for advising people of the right, such as mentioning the existence of the provision on government application forms or in pamphlets.

So, as David Burnham noted in a separate newspaper account, most Americans are unaware of this important Privacy Act provision, and the number of inspections has been minimal, despite the fact that millions of records dealing with individuals are held by such agencies as the Department of Defense, the Social Security Administration, the Department of Health and Human Services, and the Internal Revenue Service. Burnham reports, however, that the White House Office of Budget and Management, which is responsible for overseeing the act, has decided to look at government forms and brochures to make them more informative about citizen rights.

Critics of the record inspection provision contend that it is flawed in that it applies only to federal agencies and not to federally financed programs administered by the states. This means, according to Burnham, that, while Americans are entitled to see the records held by the Internal Revenue Service or the Social Security Administration, they have no such right to examine records involving food stamps or Aid to Families with

Dependent Children, federal programs that are operated by the states. The inspection provision says specifically that citizens have a right to see their records, to obtain a copy of them, and to request corrections. Agencies must promptly make the corrections or tell the individual the reason for refusal and how the decision can be appealed.[22]

Another important provision of the Privacy Act prohibits federal agencies from creating secret record systems. The law also prohibits agencies from using information acquired for one purpose for another unless they inform the person who is mentioned.

Inevitable conflicts between the Privacy Act and the Freedom of Information Act were mentioned in the previous chapter, so there is no need to repeat them here. What bears repeating, however, is to acknowledge the growing concern over both pieces of legislation. Individual journalists and journalistic organizations have noted, in the words of a spokesperson for the American Newspaper Publishers Association, "that the push for privacy not be extended to infringe upon the First Amendment right of the press to gather and disseminate news to the public." Or, as Fred Graham, CBS law correspondent, noticed, the Privacy Act has driven a wedge between the press and civil libertarians who press for greater privacy protection. This view found support from Professor Linowes, chairman of the Privacy Protection Commission, who said, "Privacy is getting a bad name in some circles simply because it is being confused with secrecy." In some cities, reporters have been denied access to police arrest records, and in some states local officials were refused essential Social Security information to determine welfare payments.[23]

There is another way to assess the conflict between public openness and informational privacy, suggested by communication law textbook authors Donald M. Gillmor and Jerome A. Barron. They remind us that the purpose of the FOIA was to increase *public* access to government-held information, whereas the Privacy Act was created to provide *individuals* more control over the gathering, dissemination, and accuracy of information about them in government files: "The latter is an FOIA for the individual." Central to both laws is the individual's role in having some say in what data are collected and who will see them. The Privacy Act does not say "no access," although some zealots would like that to be the law. The FOIA does not say "no privacy allowed," although, again, there are those in positions of power who believe the act restricts the efficient and effective operation of government and threatens privacy. Ideally, neither law should be a barrier against the other, and open *access* by the public and the individual is what's most important, with the government minimally involved—perhaps as archivist or librarian.

Surveillance of any kind—by the government, police, private corporations, the press, or the neighborhood snoop—inevitably collides with personal privacy. How much privacy, or, conversely, how much publicity, can an open society tolerate? A newspaper's demand for arrest records may be motivated by an honest desire to assess the performance of a police department, but the consequences may be exposure of individual third-party transgressions. Gillmor and Barron write: "It is important to know when denials of disclosure are based on the long tradition of official secrecy and suppression of information and when they are based on a genuine concern for a legal or constitutional right of personal privacy."[24]

Although much of the data others acquire about us may indeed be personal, not all may be deemed private in the sense that no one else may be so characterized by the information. It may be unfortunate, but nonetheless true, that, in the aggregate, individuals are more similar than different, that individualism is really a means of collective identification, not the individual as a unique and separate body. Privacy gets its strength from the community, from the others with whom the individual interacts. Access to private matter is not the same as access to the person about whom we seemingly have unique data. But even data are seldom unique and almost always used in the aggregate. Uniqueness does not come from information, nor does access to that information constitute, on its face, an invasion of personal privacy. There may be more cause for alarm if the individual's corrective right of access is withheld, or, what's worse, the individual fails to exercise the power he or she has already.

Federal legislators were aware that the Privacy Act would have the practical effect of emasculating the FOIA unless they included a provision allowing the FOIA to remain effective. During the drafting of the bill that became the Privacy Act, the Senate Government Operations Committee stated that, whereas the committee intended to implement the guarantees of individual privacy, it also planned to make available to the press and public all possible information concerning government operations in order to prevent secret databanks and unauthorized investigations.

The original House bill contained no exception for required FOIA disclosure, but, because Congress ultimately adopted the FOIA exception, the lawmakers thus expressed a commitment to the policy of disclosure stipulated in the FOIA. The Privacy Protection Commission said in its 1977 report that individual records should be withheld if disclosure would constitute a clearly unwarranted invasion, but noted also that the reverse is equally important: that records should be open if they do *not* constitute an invasion of personal privacy. It is left to the courts to decide, but so far the advantage has gone to public openness.

As mentioned throughout this chapter, a major source of the conflict between privacy and publicity, from a social as well as a legal frame of reference, is the matter of terminology. If information privacy interests are worthy of protection, the main problem would seem to be in defining—in light of the various privacy interests that may be at issue—the "intimate" or "personal" data that the individual has a right to protect. Once agreement is reached, a fair and equitable procedure could be established for deciding when a person has "waived" her or his right by making the information public. Agreement would also identify those occasions when government has a legitimate interest in the information, and it should devise safeguards for confidentiality to allow for appropriate disclosure while protecting other data from inappropriate disclosure.[25]

Notes

CHAPTER 1. In Search of Solitude

1. *The Holy Bible, American Standard Version* (New York: Thomas Nelson and Sons, 1929), Genesis 3:7.
2. Genesis 3:21.
3. John Curtis Raines, *Attack on Privacy* (Valley Forge: Judson Press, 1974), 23.
4. Arnold Stein, *Answerable Style: Essays on Paradise Lost* (Minneapolis: Univ. of Minnesota Press, 1953), 54.
5. *The Student's Milton*, ed. Frank Allen Patterson (New York: Appleton-Century-Crofts, 1933), *Paradise Lost*, IV, 220ff.
6. *Paradise Lost*, XII, 646ff.
7. Genesis 9:20–23.
8. Barrington Moore, Jr., *Privacy: Studies in Social and Cultural History* (Armonk, N.Y.: M.E. Sharpe, 1984), 267–77.
9. Ibid., 283.
10. Ibid.
11. Raymond Williams, *Keywords: A Vocabulary of Culture and Society* (New York: Oxford Univ. Press, 1976), 203.
12. Ibid., 204.
13. David H. Flaherty, *Privacy in Colonial New England* (Charlottesville: Univ. Press of Virginia, 1972), 26.
14. Ibid., 30.
15. J. C. Furnas, *The Americans: A Social History of the United States, 1587–1914* (New York: G. P. Putnam's Sons, 1969), 164.
16. Flaherty, 39.
17. Ibid., 44.
18. Ibid., 55.
19. Moore, 67.

20. Furnas, 164.
21. Flaherty, 92.
22. Thomas H. O'Connor, "The Right to Privacy in Historical Perspective," 53 *Mass. L. Quarterly* 101 (1968) at 101–2.
23. *The Civil Ruler . . . A Serman Preached Before the General Assembly of the Colony of Connecticut . . .* (New London: 1753), 49; cited in Michael Kammen, *People of Paradox: An Inquiry Concerning the Origins of American Civilization* (New York: Knopf, 1973), 93.
24. Cited by Flaherty, 87.
25. Ibid., 87–88.
26. Ibid., 90.
27. Ibid., 96.
28. Ibid.
29. Ibid., 97.
30. Harvey L. Zuckerman and Martin J. Gaynes, *Mass Communication Law in a Nutshell* (St. Paul: West Publishing, 1977), 81.
31. Flaherty, 106.
32. Paul Weiss, *Privacy* (Carbondale: Southern Illinois Univ. Press, 1983), 299.
33. Raines, p. 16.
34. Margaret Drabble, *The Ice Age* (New York: Popular Library Edition, 1977), 45.
35. Flaherty, 110.
36. *The Student's Milton, Samson Agonistes*, 865ff.
37. O'Connor, 105.
38. Ibid.
39. Gay Wilson Allen, *Waldo Emerson: A Biography* (New York: Viking, 1981), 373–74, 641, 652.
40. *Walden* (Princeton: Princeton Univ. Press, 1973), 130.
41. Ibid., 117–18.
42. Bernard DeVoto, *The Year of Decision: 1846* (Boston: Little, Brown, 1943), 37.
43. O'Connor, 107.
44. Henry James, *The Portrait of a Lady*, Vol. 1 (Boston: Houghton Mifflin, 1881), 287–88.
45. Ibid., 250.
46. Ibid., 287–88.
47. *The Reverberator* (New York: Macmillan, 1888), 67–68.
48. Ibid., 134–35.
49. Ibid., 172, 179–80.
50. bid., 209, 215.
51. O'Connor, 109.

CHAPTER 2. Creating a Legal Right

1. Esther Forbes, *Paul Revere: The World He Lived In* (Boston: Houghton Mifflin, 1942), 70.

2. Edward Shils, "Privacy: Its Constitution and Vicissitudes," 31 *Law and Contemporary Problems* 281 (1966) at 290.

3. Ibid., 290–91.

4. Thomas H. O'Connor, "The Right to Privacy in Historical Perspective," 53 *Mass. L. Quarterly* 101 (1968) at 109.

5. Don R. Pember, *Privacy and the Press: The Law, the Mass Media, and the First Amendment* (Seattle: Univ. of Washington Press, 1972), 10–14.

6. "The Professor of Journalism," *The Nation,* July 17, 1873, p. 37.

7. "Newspaper Espionage," *Forum*, Aug. 1886, p. 533.

8. "The Rights of the Citizen, IV: To His Own Reputation," July 1890, p. 67.

9. "The Right to Privacy," 4 *Harvard L. R.* 193 (1890), Samuel D. Warren and Louis D. Brandeis.

10. Pember, 24; Alpheus Thomas Mason, *Brandeis: A Free Man's Life* (New York: Viking, 1946), 70; and Allan Nevins, *American Press Opinion* (New York: Heath, 1928), 299.

11. Thomas M. Cooley, *A Treatise on the Law of Torts or the Wrongs Which Arise Independently of Contract*, 4th ed. by D. Avery Haggard, Vol. 1 (Chicago: Callaghan, 1932), 34.

12. Cited in Alan F. Westin, *Privacy and Freedom* (New York: Atheneum, 1970), 345.

13. Ibid., 345–46.

14. Cooley, 34.

15. Pember, esp. Chap. 3.

16. Diane L. Zimmerman, "Requiem for a Heavyweight: A Farewell to Warren and Brandeis Privacy Tort," 68 *Cornell L. R.* 291 (1983) at 294.

17. Milton R. Konvitz, "Privacy and the Law: A Philosophical Prelude," 31 *Law and Contemporary Problems* 272 (1966) at 279–80.

18. Note, "The Right to Privacy in Nineteenth Century America," 94 *Harvard L. R.* 1892 (1981) at 1894–96.

19. Note at 1899.

20. J. C. Furnas, *The Americans: A Social History of the United States, 1587–1914* (New York: G. P. Putnam's Sons, 1969), 356.

21. Note at 1901.

22. Note at 1902–3.

23. Note at 1904–5.

24. Roscoe Pound, "The Fourteenth Amendment and the Right of Privacy," 13 *Western Reserve L. R.* 34 (1961) cited at 37.

25. *Schuyler v. Curtis*, 15 N.Y. Supp. 787 (1892) at 788.

26. 171 N.Y. 538 (1902).

27. Richard F. Hixson, "Whose Life Is It, Anyway? Information as Property," *Information and Behavior*, ed. Brent D. Ruben, Vol. 1 (New Brunswick: Transaction Books, 1985), 78.

28. Denis O'Brien, "The Right of Privacy," 2 *Columbia L. R.* 437 (1902) at 437–48.

29. William Prosser, *Handbook of the Law of Torts*, 4th ed. (St. Paul: West Publishing, 1971) at 802–3.

30. 122 Ga. 190 (1905).

31. at 217–18.
32. Zimmerman, 365.
33. William E. François, *Mass Media Law and Regulation*, 3d ed. (Columbus: Grid Publishing, 1982), 197.
34. *Sweenek v. Pathé News Inc.*, 16 F. Supp. 746 (1936) at 747.
35. "James Madison and the Constitution," *The Wilson Quarterly*, Vol IX, No. 3, Summer 1985, p. 91.
36. Frederick Davis, "What Do We Mean by 'Right to Privacy'?" 4 *South Dakota L. R.* 1 (1959) at 6.
37. Pember, 118.
38. Davis, 18, 20, 23.
39. Alan R. White, *Rights* (Oxford: Clarendon Press, 1984), 2.
40. Ibid., 16.
41. Ibid., 175.
42. Richard E. Flathman, *The Practice of Rights* (Cambridge: Cambridge Univ. Press, 1976), 34.
43. Ibid.
44. Samuel J. Konefsky, *The Legacy of Holmes and Brandeis: A Study in the Influence of Ideas* (New York: DaCapo, 1974), 263–64.
45. O'Connor, 113.

CHAPTER 3. False Promises, Myriad Objectives

1. J. B. Young, ed., *Privacy* (New York: Wiley, 1978), esp. "Introduction: A Look at Privacy."
2. Raymond Wacks, "The Poverty of Privacy," 96 *Law Quarterly Review* 73 (1980) at 77.
3. William L. Prosser, "Privacy," 48 *Calif. L. R.* 383 (1960).
4. Edward J. Bloustein, "Privacy—An Aspect of Human Dignity: An Answer to Dean Prosser," 39 *New York University L. R.* 962 (1964).
5. Alan F. Westin, *Privacy and Freedom* (New York: Atheneum, 1970).
6. Executive Office of the President, Office of Science and Technology, *Privacy and Behavioral Research* (Washington, D.C.: GPO, 1967).
7. *Handbook of the Law of Torts* (St. Paul: West Publishing, 1955). A second edition of the *Handbook* appeared in 1963. In 1954, Prosser undertook to serve as reporter for *Restatement*, published by the American Law Institute and printed by West Publishing of St. Paul. The second edition, supervised by Prosser, began to appear in 1965. Prosser resigned from that post in 1970. He died in 1972.
8. Diane L. Zimmerman, "Requiem for a Heavyweight: A Farewell to Warren and Brandeis's Privacy Tort," 68 *Cornell L. R.* 291 (1983) at 333.
9. Westin, 31.
10. Ibid.
11. Cited in ibid.
12. Ibid., 32.
13. Thomas I. Emerson, *Toward a General Theory of the First Amendment* (New York: Random House, 1966).

14. *Palko v. Connecticut*, 302 U.S. 319, 327 (1937).

15. Westin, 33.

16. Ibid., 35.

17. Ralph Waldo Emerson, "Friendship," in *The Works of Ralph Waldo Emerson* (Boston: Houghton Mifflin, 1888), II, 193.

18. Cited by Westin, 34.

19. *The Complete Poetry and Selected Prose of John Donne*, ed. Charles M. Coffin (New York: Modern Library, 1952), 191.

20. G. W. F. Hegel, *Philosophy of Right*, trans. T. M. Knox (Oxford: Clarendon, 1967), Part 187.

21. Gary L. Bostwick, "A Taxonomy of Privacy: Repose, Sanctuary and Intimate Decision," 64 *Calif. L R.* 1447 (1976).

22. R. Allan Dionisopoulos and Craig R. Ducat, *The Right to Privacy: Essays and Cases* (St. Paul: West Publishing, 1976).

23. Bostwick, 1448.

24. In order: *Lehman v. Shaker Heights*, 418 U.S. 298 (1974); *In re Quinlan*, 70 N.J. 10 (1976); *Doe v. Commonwealth's Attorney*, 403 F. Supp. 1199 (1975); and *Belle Terre v. Borass*, 416 U.S. 1 (1974).

25. Bostwick, 1451.

26. Ibid., 1452.

27. Ibid.

28. *Katz v. U.S.*, 389 U.S. 347 (1967); *U.S. v. White*, 401 U.S. 745 (1971); and *Time Inc. v. Hill*, 385 U.S. 374 (1967).

29. *Griswold v. Connecticut*, 381 U.S. 479 (1965).

30. Bostwick, 1482.

31. Judee K. Burgoon, "Privacy and Communication," *Communication Yearbook 6*, ed. Michael Burgoon (Beverly Hills: Sage, 1982).

32. Ibid., 219.

33. P. A. Kelvin, "Social Psychological Examination of Privacy," 12 *British Journal of Social and Clinical Psychology* 248 (1973) at 251.

34. *The Sociology of Georg Simmel*, ed. and trans. K. H. Wolff (New York: Macmillan, 1950), 322.

35. J. M. Caroll, *Confidential Information Sources: Public and Private* (Los Angeles: Security World, 1975), 277.

36. Council of State Governments, *Privacy: A Summary of a Seminar on Privacy* (Lexington: Author, 1975), 41.

37. Burgoon, 232.

38. Louis Lusky, "Invasion of Privacy: A Clarification of Concepts," 72 *Columbia L. R.* 693 (1972).

39. Wacks, 88.

40. Edward T. Hall, *The Hidden Dimension* (Garden City, N.Y.: Doubleday, 1966) and *The Silent Language* (Garden City, N.Y.: Doubleday, 1973).

CHAPTER 4. Disagreement on Zones

1. 116 U.S. 616 (1886).

2. 141 U.S. 250 (1891).

3. 262 U.S. 390 (1923).

4. 268 U.S. 510 (1925).

5. 316 U.S. 535 (1942).

6. 357 U.S. 449 (1958).

7. 277 U.S. 438 (1928).

8. 367 U.S. 539 (1961).

9. In *Katz v. United States*, 389 U.S. 347 (1967), the Supreme Court held that wiretapping, as well as electronic surveillance, was included among protections of the Fourth Amendment and that the use of such methods by law enforcement agencies must conform to the requirements of that constitutional provision.

10. 407 U.S. 297 (1972).

11. *New York Times*, June 9, 1965, p. 46.

12. Fred W. Friendly and Martha J. H. Elliott, *The Constitution: That Delicate Balance* (New York: Random House, 1984), 189.

13. Ibid., 199.

14. John Lukacs, *Outgrowing Democracy: A History of the United States in the Twentieth Century* (New York: Doubleday, 1984), 173.

15. Michael Kammen, ed., *The Contrapuntal Civilization: Essays Toward a New Understanding of the American Experience* (New York: Crowell, 1971), 26.

16. Lawrence Lader, *Abortion* (New York: Bobbs-Merrill, 1966), 88.

17. Cited in R. Sauer, "Attitudes Toward Abortion in America, 1800–1973," 28 *Population Studies* 53 (1974).

18. *Roe v. Wade*, 410 U.S. 113 (1973) and *Doe v. Bolton*, 410 U.S. 179 (1973), decided together.

19. Brian Berry, "Courts and Constitutions," in the *Times Literary Supplement*, Oct. 25, 1985, p. 1195.

20. See Kristin Luker, *Abortion and the Politics of Motherhood* (Berkeley: Univ. of California Press, 1984).

21. Ibid., 141.

22. John T. Noonan, Jr., *A Private Choice: Abortion in America in the Seventies* (New York: Free Press, 1979), 1–2.

23. Ibid., 79.

24. Ibid., 190, 192.

25. John Hart Ely, "The Wages of Crying Wolf: A Comment on *Roe v. Wade*," 82 *Yale L. J.* 920 (1973) at 947.

26. Paul Bender, "Privacies of Life," *Harper's*, April 1974, pp. 41–44.

CHAPTER 5. Rationale of Public

1. James Steintrager, *Bentham* (Ithaca: Cornell Univ. Press, 1977), 58.

2. *The Complete Works of Jeremy Bentham*, ed. John Bowring (Edinburgh: William Tait, 1843). Cited throughout as *Works*.

3. Mary P. Mack, *Jeremy Bentham: An Odyssey of Ideas, 1748–1792* (London: Heinemann, 1962), 190.

4. L. W. Sumner, "Rights Denaturalized," in *Utility and Rights*, ed. R. G. Frey (Minneapolis: Univ. of Minnesota Press, 1984), 32.

5. Nancy L. Rosenblum, *Bentham's Theory of the Modern State* (Cambridge: Harvard Univ. Press, 1978), 153.

6. John Stuart Mill, *On Liberty*, ed. Currin V. Shields (Indianapolis: Bobbs-Merrill, 1956), xvii.

7. A. C. Dicey, *Lectures on the Relations Between Law & Public Opinion in England During the Nineteenth Century* (London: Macmillan, 1905), 136.

8. Iredell Jenkins, *Social Order and the Limits of Law: A Theoretical Essay* (Princeton: Princeton Univ. Press, 1980).

9. Richard H. Rovere, "The Invasion of Privacy (1): Technology and the Claims of Community," 27 *American Scholar* 413 (1958), 419.

10. John Stuart Mill, *On Liberty*, 82.

11. Cited in Frey, 129.

12. Jeremy Waldron, "Rights and Trade-Offs," *Times Literary Supplement*, Nov. 8, 1985, pp. 1269–70.

13. Cited in ibid., 1269.

14. *Works*, II, 501.

15. Jenkins, 242.

16. Ibid., 244.

17. Ibid., 246.

18. Ibid., 247.

19. Ibid., 252.

20. Hyman Gross, "Privacy and Autonomy," in *NOMOS XIII Privacy*, eds. J. Roland Pennock and John W. Chapman (New York: Atherton, 1971), 176, 181.

21. Ruth Gavison, "Information Control: Availability and Exclusion," in *Public and Private in Social Life*, eds. S. I. Benn and G. F. Gaus (New York: St. Martin's Press, 1983), 124.

22. Jenkins, 256.

23. James Griffin, "Towards a Substantive Theory of Rights," in *Utility and Rights*, 156.

24. Michael Kammen, *People of Paradox* (New York: Knopf, 1973), 292.

25. Griffin, 145.

26. *Works*, VI, 28.

27. Ibid., 351–80.

28. Ibid., 357.

29. Ibid., 367.

30. Ibid., 369–70.

31. Ibid., 372.

32. Robert Nisbet, *Twilight of Authority* (New York: Oxford Univ. Press, 1975), 251–52.

33. Steintrager, 121.

34. Ibid., 125.

35. Ronald Dworkin, *Taking Rights Seriously* (Cambridge: Harvard Univ. Press, 1977).

36. Ibid., 81, 84.

37. Dworkin, "The Rights of Myron Farber," *New York Review of Books*, Oct. 26, 1978, p. 34.

38. Dworkin, "Is the Press Losing the First Amendment?", *New York Review of Books*, Dec. 4, 1980, p. 51.

39. Griffin, 152, 158.

40. H. L. A. Hart, "Between Utility and Rights," in Alan Ryan, ed., *The Idea of Freedom: Essays in Honour of Isaiah Berlin* (Oxford: Oxford Univ. Press, 1979), 81.

41. Ibid., 86–87.

CHAPTER 6. A Sense of Community

1. From "Song of Myself," Verse 51, in *Walt Whitman: Complete Poetry and Selected Prose*, ed. James E. Miller, Jr. (Boston: Houghton Mifflin, 1959), 68.

2. Erik Erikson, *Childhood and Society*, 2d ed. (New York: Norton, 1963), 285–86.

3. E. Douglas Branch, *The Sentimental Years, 1836–1860* (New York: D. Appleton-Century, 1934), 16.

4. Emerson cited in Warren I. Susman, *Culture as History: The Transformation of American Society in the Twentieth Century* (New York: Pantheon, 1984), 213; also see Michael Kammen, *People of Paradox* (New York: Knopf, 1973), 291.

5. Susman, 212–13.

6. Kammen, 108.

7. Cited in Ibid., 108–9.

8. Ibid., pp. 109–10.

9. Jacques Maritain, *Reflections on America* (New York: Scribner's, 1958), 163.

10. "Community and Equality in Conflict," *New York Times*, Sept. 8, 1985, p. E25.

11. Robert N. Bellah et al., *Habits of the Heart: Individualism and Commitment in American Life* (Berkeley: Univ. of California Press, 1985), 286.

12. Raymond Williams, *Keywords: A Vocabulary of Culture and Society* (New York: Oxford Univ. Press, 1976), 65.

13. Ibid., 66.

14. Ibid., 135.

15. Barrington Moore, Jr., *Privacy: Studies in Social and Cultural History* (Armonk: M. E. Sharpe, 1984), 5.

16. Ibid., 12.

17. Hannah Arendt, *The Human Condition* (Chicago: Univ. of Chicago Press, 1958), p. 38.

18. Ibid.

19. Ibid.

20. Ibid.

21. Nicholas Gage, *Eleni* (New York: Ballantine, 1984), 41–42.

22. Moore, 118.

23. John Lukacs, *Outgrowing Democracy: A History of the United States in the Twentieth Century* (New York: Doubleday, 1984), 177.

24. Ibid., 379–80.

25. David H. Flaherty, *Privacy in Colonial New England* (Charlottesville: Univ. Press of Virginia, 1972), 188.

26. Robert Frost, "Mending Wall," 1914. (Although Frost was not there at the time, his kind of irony was!)

27. Andrew Jackson Downing, *Rural Essays*, ed. George William Curtis (New

York: DaCapo Press, 1974), 142. A republication of the first edition, published in 1853.

28. Ibid., 142.

29. *Democracy in America,* ed. J. P. Mayer (New York: Doubleday, 1969), II, 506.

30. Ibid., 525–28.

31. *Sybil: or The Two Nations* (Oxford: Oxford Univ. Press, 1845; World's Classics Edition, 1926), 197.

32. *New York Times Book Review,* April 14, 1985, pp. 1, 22.

33. Warren Johnson, *The Future Is Not What It Used To Be: Returning to Traditional Values in an Age of Scarcity* (New York: Dodd, Mead, 1985).

34. Arendt, 50–52.

35. Ibid., 50.

36. Ibid., 52–53.

37. Ibid., 57.

38. Ibid., 58.

39. Raymond Williams, *The Country and the City* (New York: Oxford Univ. Press, 1973), 295.

40. Richard D. Brown, *Modernization: The Transformation of American Life, 1600–1865* (New York: Hill and Wang, 1976), 195.

41. Richard H. Rovere, "The Invasion of Privacy (1): Technology and the Claims of Community," 27 *The American Scholar* 413 (1958), 421.

CHAPTER 7. Copyright of Personality

1. Tamar Lewin, "Whose Life Is It, Anyway? Legally It's Hard to Tell," *New York Times,* Nov. 21, 1982, Sec. 2, pp. 1, 26.

2. Raymond Williams, *Keywords: A Vocabulary of Culture and Society* (New York: Oxford Univ. Press, 1976).

3. Ibid., 195–97.

4. Warren I. Susman, *Culture as History: The Transformation of American Society in the Twentieth Century* (New York: Pantheon, 1984), 272.

5. Ibid., 274.

6. Cited in ibid.

7. Ibid., 275–76.

8. Ibid., 277.

9. Philip Rieff, *Freud: The Mind of a Moralist* (New York: Doubleday, 1961), 391–92. See esp. chap. 10, "The Emergence of Psychological Man."

10. Susman, 280.

11. Cited in ibid., 281.

12. Richard Schickel, *His Picture in the Papers: A Speculation on Celebrity in America Based on the Life of Douglas Fairbanks, Sr.* (New York: Charterhouse, 1974), 9.

13. 202 F. 2d 866 (1953) at 868.

14. Erik D. Lazar, "Towards a Right of Biography: Controlling Commercial Exploitation of Personal History," 2 *Comm/Ent L. J.* 489 (1980), at 535–36.

15. Kevin Marks, "An Assessment of the Copyright Model in Right of Publicity Cases," 70 *Calif. L. R.* 786 (1982) at 789.

16. 433 U.S. 562 (1977).

17. 351 N.E. 2d, 454, 462 (1972) at 460.

18. at 460.

19. at 466.

20. 53 L. Ed. 2d 965 (1977) at 975.

21. at 975–76.

22. 433 U.S. 562 at 579.

23. at 583.

24. Bennett D. Zurofsky, "Constitutional Law—Privacy Torts—First Amendment Does Not Privilege Violation of Right of Publicity—*Zacchini v. Scripps-Howard Broadcasting Co.*," 31 *Rutgers L. R.* 269 (1978) at 294.

25. Zurofsky at 295.

26. Alfred McClung Lee to author, Jan. 15, 1986.

27. Richard Posner, "The Right to Privacy," 12 *Georgia L. R.* 393 (1978).

28. Posner, 12 *Georgia L. R.* 393 at 403.

29. "Privacy Is Dear at Any Price: A Response to Professor Posner's Economic Theory," 12 *Georgia L. R.* 429 (1978) at 453.

30. at 407.

31. George Steiner, "Language under Surveillance: The Writer and the State," *New York Times Book Review*, Jan. 12, 1986, p. 36.

32. Barbara Goldsmith, "The Meaning of Celebrity," *The New York Times Magazine*, Dec. 4, 1983, p. 75.

33. Alexander Meiklejohn, *Political Freedom* (New York: Harper and Row, 1960), 8–9.

34. Zechariah Chafee, *Government and Mass Communications: A Report from the Commission on Freedom of the Press* (Chicago: Univ. of Chicago Press, 1947), 138.

35. Roscoe Pound, "Interests of Personality, Part 1," 28 *Harvard L. R.* 343, 363 (1915).

36. Cited in William E. François, *Mass Media Law and Regulation*, 4th ed. (New York: Wiley, 1986), 616.

37. Marc J. Apfelbaum, "Copyright and the Right of Publicity: One Pea in Two Pods," 71 *Georgetown L. J.* 1567 (1983), at 1567.

38. Apfelbaum at 1570.

39. Lewin, "Whose Life Is It Anyway?", 26.

40. *Rosemont Enterprises Inc. v. Random House*, 366 F. 2d 303 (2d Cir. 1966); cited in Francois, 4th ed., 616ff.

41. Steiner, *New York Times Book Review*, Jan. 12, 1986, p. 36.

CHAPTER 8. Beware the Watchdog

1. Alexander Meiklejohn, "The First Amendment Is an Absolute," *Supreme Court Review* (1961), 255; J. Skelly Wright, "Defamation, Privacy, and the Public's Right to Know: A National Problem and a New Approach," 46 *Texas L. R.* 630 (1968) at 633.

2. George Steiner, "Language under Surveillance: The Writer and the State," *New York Times Book Review*, Jan. 12, 1986, p. 36.

3. Arnold E. Lubasch, "$10 Million Invasion-of-Privacy Award Upset," *New York Times*, Sept. 16, 1984, p. 55.

4. Stuart Taylor, Jr., "Life in the Spotlight: Agony of Getting Burned," *New York Times*, Feb. 27, 1985, p. A16.

5. J. H. Plumb, "Private Lives, Public Faces," 16 *Horizon* (Spring 1974), 56–57.

6. "Text of Resignation Letter," *New York Times*, Feb. 27, 1985, p. D5.

7. Plumb, 57.

8. "On Private Transgressions and Holding the Public Trust," *New York Times*, March 3, 1985, p. 2E.

9. *New York Times*, Feb. 27, 1985, p. A16.

10. Ibid.

11. Plumb, 57.

12. Richard Kurnit, "The 'Miss America' Pictures Controversy and the Right to Privacy," *New York L. J.*, Aug. 3, 1984, p. 5.

13. *Arrington v. New York Times Co.*, 55 N.Y. 2d 433 (1982), cert. denied 103 S.Ct. 787 (1983).

14. *Shields v. Gross*, 58 N.Y. 2d 338 (1983); *Flores v. Mosler Safe Co.*, 7 N.Y. 2d 276 (1959); *Namath v. Sports Illustrated*, 48 A.D. 2d 487 (1975); *Paulsen v. Personality Posters Inc.*, 59 Misc. 2d 444 (1968).

15. *Time Inc. v. Hill*, 382 U.S. 374 at 388.

16. Kurnit, 5.

17. Wright at 632.

18. Cited in Victor A. Kovner, "Recent Developments in Intrusion, Private Facts, False Light, and Commercialization Claims," *Communication Law 1984* (New York: Pracfising Law Institute, 1984), 425.

19. Cited in Don R. Pember, *Mass Media Law*, 2d ed. (Dubuque: Wm. C. Brown, 1981), 239.

20. *Cassidy v. ABC*, 60 Ill. App. 3d 831.

21. Pember, 241.

22. 420 U.S. 469 (1975).

23. Mark Schadrack, "Privacy and the Press: A Necessary Tension," 18 *Loyola of Los Angeles L. R.* 949 (1985) at 952.

24. Gerald G. Ashdown, "Media Reporting and Privacy Claims—Decline in Constitutional Protection for the Press," 66 *Kentucky L. J.* 761 (1977–78) at 770.

25. 420 U.S. 469 at 487.

26. Ashdown at 774.

27. *Gertz v. Welch*, 418 U.S. 323 (1974) at 346.

28. 698 F. 2d 831.

29. Ashdown at 781.

30. *Time Inc. v. Hill* at 387–88.

31. Melville B. Nimmer, "The Right to Speak from *Times* to *Time*: First Amendment Theory Applied to Libel and Misapplied to Privacy," 56 *Calif. L. R.* 935 (1968).

32. *Sharon v. Time Inc.*, 599 F. Supp. 538 (S.D.N.Y. 1984); *Westmoreland v. CBS*, 10 Med. L. Rptr. (BNA) 2417 (S.D.N.Y. 1984).

33. *Curtis Publishing Co. v. Butts* and *Associated Press v. Walker*, 388 U.S. 130 (1967); *Rosenblum v. Metromedia Inc.*, 403 U.S. 29 (1971).

34. *Time Inc. v. Firestone*, 424 U.S. 448 (1976); *James v. Gannett Co. Inc.*, 386 N.Y.S. 2d 871 (1976).

35. Marilyn A. Lashner, "Privacy and the Public's Right to Know," 53 *Journalism Quarterly* 679 (1976) at 688.

36. Gerald J. Baldesty and Roger A. Simpson, "The Deceptive 'Right to Know': How Pessimism Rewrote the First Amendment," 56 *Washington L. R.* 366 (1981) at 365.

37. *Whitney v. California*, 274 U.S. 357.

38. Norman Dorsen, Paul Bender, and Burt Neuborne, *Emerson, Haber, and Dorsen's Political and Civil Rights in the United States*, 4th ed., Vol. 1 (Boston: Little, Brown, 1976), 540.

39. *Hunter v. Washington Post*, 43 U.S.L.W. 2059 (1974); *Vogel v. W. T. Grant Co.*, 327 A.2d 133 (1974).

40. *Seattle Times v. Rhinehart*, 81 L. Ed. 2d 17 (1984).

41. *Richmond Newspapers Inc. v. Virginia*, 448 U.S. 555 (1980) at 638.

42. *Press-Enterprise v. Superior Court*, 78 L. Ed. 2d (1984) at 636.

43. *Press-Enterprise* at 638.

44. Dan Paul, Richard J. Ovelman, James D. Spaniolo, and Steven M. Kamp, "Access after *Press-Enterprise*," *Communication Law 1984* (New York: Practising Law Institute, 1984), 47.

45. 724 F. 2d 1010, vacated and remanded, 737 F. 2d 1170 (1984).

46. *Zemel v. Rusk*, 381 U.S. 1 (1965).

47. *Pell v. Procunier*, 417 U.S. 817 (1974); *Saxbe v. Washington Post Co.*, 417 U.S. 843 (1974).

48. *Briscoe v. Reader's Digest Association*, 4 Cal. 3d 529, 93 Cal. Rptr. 866 (1971) at 869.

CHAPTER 9. Privatizing Information

1. Diane L. Zimmerman, "Requiem for a Heavyweight: A Farewell to Warren and Brandeis' Privacy Tort," 68 *Cornell L. R.* 291 (1984) at 362–63.

2. Zimmerman at 363.

3. "The Right to Be Let Alone," 15 *Center Magazine* 31 (Sept./Oct. 1982) at 33.

4. at 33.

5. at 34.

6. "The Freedom of Information Act's Privacy Exemption and the Privacy Act of 1974," 11 *Harvard Civil Rights Civil Liberties L. R.* 596 (1976) at 627.

7. Marc Arnold and Andrew Kisseloff, "An Introduction to The Federal Privacy Act of 1974 and Its Effect on the Freedom of Information Act," 11 *New England L. R.* 463 (1976) at 482.

8. Letter to W. T. Barry, Aug. 4, 1882, *The Writings of James Madison*, ed. Gaillard Hunt (New York: G.P. Putnam's Sons, 1910), Vol. IX, p. 103. Cited in *Environmental Protection Agency v. Mink*, 410 U.S. 73 (1973) at 110–11.

9. 441 U.S. 281 (1979).

10. Kimera Maxwell and Roger Reinsch, "The Freedom of Information Act Privacy Exemption: Who Does It Really Protect?" 7 *Communication and the Law* 45 (April 1985) at 45.

11. John Ullman and Karen List, "An Analysis of Government Cost Estimates of Freedom of Information Act Compliance," 62 *Journalism Quarterly* 465 (Autumn 1985) at 465.

12. Cited in Marie Veronica O'Connell, "A Control Test for Determining 'Agency Record' Status Under the Freedom of Information Act," 85 *Columbia L. R.* 611 (1985) at 621.

13. Steve Weinberg, "Freedom of Information: You Still Need a Can Opener," 220 *The Nation* 463 (April 19, 1975) at 464.

14. 445 U.S. 169, 185–86 (1980).

15. 445 U.S. 136 (1980).

16. Donald M. Gillmore and Jerome A. Barron, *Mass Communication Law: Cases and Comment*, 4th ed. (St. Paul: West Publishing, 1984), 441.

17. Ibid., 441.

18. O'Connell, 612.

19. "Freedom of Information Act Faces Crippling Attacks," *The Courier-News*, Bridgewater, N.J., Sept. 15, 1981, A7.

20. Floyd Abrams, "The New Effort to Control Information," *New York Times Magazine*, Sept. 25, 1983, p. 72.

21. Ibid., 72.

22. 425 U.S. 352 (1976).

23. at 372.

24. *National Parks and Conservation Association v. Morton*, 498 F. 2d 765 (D.C. Cir. 1974).

25. Richard B. Kielbowicz, "The Freedom of Information Act and Government's Corporate Information Files," 55 *Journalism Quarterly* 481, 526 (Autumn 1978) at 483.

26. 478 F. Supp. 103 (D.D.C. 1979).

27. Kevin Condrin Dwyer and Peter Glatz Rush, "Developments Under the Freedom of Information Act—1983," 1984 *Duke L. J.* 377 (1984) footnote at 405.

28. Kielbowicz at 486.

29. *NLRB v. Sears, Roebuck & Co.*, 421 U.S. 132 (1975).

30. Gillmor and Barron, 459.

31. *Niemeier v. Watergate Special Prosecution Force*, 3 Med. L. Rpt. 1321, 565 F. 2d 967 (7th Cir. 1977).

32. *Sims v. CIA*, 642 F. 2d 569 (1980).

33. *Getman v. NLRB*, 450 F. 2d 670 (1971).

34. Cited in Dwyer and Rush, *Duke L. J.* at 415.

35. *Robles v. EPA*, 484 F. 2d 843 (4th Cir. 1973).

36. Cited in Dwyer and Rush, *Duke L. J.* at 415.

37. Maxwell and Reinsch at 54–55.

38. 456 U.S. 615 (1982).

39. Kielbowicz at 482.

40. House Committee on Government Operations, *Hearings on U.S. Government Information Policies and Practices—Administration and Operation of the Freedom*

of Information Act, 92nd Cong., 2nd sess., 1972, pt. 6, pp. 1970–72. (correspondence between Ralph Nader and the chairman of the Federal Power Commission, Feb. 1972). Cited in Kielbowicz, footnote at 482.

41. *Pennzoil Co. v. FPC*, 534 F. 2d 627 (5th Cir. 1976).
42. Richard Neely, *How Courts Govern America* (New Haven: Yale Univ. Press, 1981), 218–19.
43. "Free Speech v. Scalia," *New York Times*, April 29, 1985, A17.
44. "Assessing the Freedom of Information Act," *New York Times*, Aug. 29, 1985, B10.

CHAPTER 10. Privacy

1. David Burnham, *The Rise of the Computer State* (New York: Vintage, 1984), 224.
2. Paul Sieghart, *Privacy and Computers* (London: Latimer, 1976), 15.
3. Alan F. Westin, *Privacy and Freedom* (New York: Atheneum, 1967), 7.
4. Arthur R. Miller, *The Assault on Privacy: Computers, Data Banks, and Dossiers* (Ann Arbor: Univ. of Michigan Press, 1971), 40.
5. Burnham, p. 11.
6. Miller, p. 180.
7. Ibid.
8. Ibid., 181.
9. "Must Personal Privacy Die in the Computer Age?" 65 *A.B.A.J.* 1180 (1979) at 1182.
10. Albert O. Hirschman, *Shifting Involvements: Private Interest and Public Action* (Princeton: Princeton Univ. Press, 1982), 126; Jacob Burckhardt, *Force and Freedom: Reflections on History* (New York: Pantheon, 1943), 118.
11. *New York Times*, Sept. 8, 1985, p. 59.
12. Miller, 181.
13. Hirschman, 122; Garry Wills, *Inventing America: Jefferson's Declaration of Independence* (New York: Doubleday, 1978), chaps. 10 and 18.
14. Walter M. Sullivan, *Reconstructing Public Philosophy* (Berkeley: Univ. of California Press, 1982), 159.
15. *The Nation*, Sept. 28, 1974, p. 260.
16. Ibid., 261.
17. Ibid.
18. 416 U.S. 21, 94 S. Ct. 1494, 38 L. Ed. 2nd 812.
19. 425 U.S. 435.
20. Joel B. Grossman and Richard S. Wells, *Supplemental Cases for Constitutional Law and Judicial Policy Making* (New York: Wiley, 1975), 259.
21. Ronald Dworkin, *Taking Rights Seriously* (Cambridge: Harvard Univ. Press, 1977), 22.
22. Burnham, *New York Times*, April 20, 1985, p. 8.
23. William E. François, *Mass Media Law and Regulation*, 3rd ed. (Columbus: Grid Publishing, 1982), 258.

24. Donald M. Gillmor and Jerome A. Barron, *Mass Communication Law: Cases and Comment*, 4th ed. (St. Paul: West Publishing, 1984), 480.

25. Norman Dorsen, Paul Bender, and Burt Neuborne, *Emerson, Haber, and Dorsen's Political and Civil Rights in the United States*, 4th ed., Vol. 1 (Boston: LIttle, Brown, 1976), 846.

Index

124708